RURAL REVIVAL

Perspectives on Rural Policy and Planning

Series Editors:
Andrew Gilg, University of Exeter and University of Gloucestershire, UK
Henry Buller, University of Exeter, UK
Owen Furuseth, University of North Carolina, USA
Mark Lapping, University of South Maine, USA

Other titles in the series

**Rural Housing, Exurbanization, and Amenity-Driven Development
Contrasting the "Haves" and the "Have Nots"**
Edited by David Marcouiller, Mark Lapping and Owen Furuseth
ISBN 978 0 7546 7050 6

**Rural Policing and Policing the Rural
A Constable Countryside?**
Edited by Rob I. Mawby and Richard Yarwood
ISBN 978 0 7546 7473 3

**Participatory Rural Planning
Exploring Evidence from Ireland**
Michael Murray
ISBN 978 0 7546 7737 6

**Naming Food After Places
Food Relocalisation and Knowledge Dynamics in Rural Development**
Edited by Maria Fonte and Apostolos G. Papadopoulos
ISBN 978 0 7546 7718 5

**A Living Countryside?
The Politics of Sustainable Development in Rural Ireland**
Edited by John McDonagh, Tony Varley and Sally Shortall
ISBN 978 0 7546 4669 3

Rural Revival?
Place Marketing, Tree Change
and Regional Migration in Australia

JOHN CONNELL and PHIL McMANUS
University of Sydney, Australia

Routledge
Taylor & Francis Group

LONDON AND NEW YORK

First published 2011 by Ashgate Publishing

2 Park Square, Milton Park, Abingdon, Oxon OX14 4RN
711 Third Avenue, New York, NY 10017, USA

Routledge is an imprint of the Taylor & Francis Group, an informa business

First issued in paperback 2016

British Library Cataloguing in Publication Data
Rural revival : place marketing, tree change and regional
migration in Australia. -- (Perspectives on rural policy
and planning)
1. Rural population--Australia. 2. Rural-urban migration--
Australia. 3. Urban-rural migration--Australia. 4. Place
marketing--Australia. 5. Country Week (Organization)
I. Series II. Connell, John, 1946- III. McManus, Phil,
1966-
304.6'2'0994'091734-dc22

Library of Congress Cataloging-in-Publication Data
Connell, John, 1946-
  Rural revival? : place marketing, tree change and regional migration in Australia / by John
Connell and Phil McManus.
    p. cm. -- (Perspectives on rural policy and planning)
  Includes index.
  ISBN 978-0-7546-7511-2 (hbk) 1. Rural-urban migra-
tion--Australia. 2. Urban-rural migration--Australia. 3. Rural renewal--Australia. 4. Rural de-
velopment--Australia. 5. Place marketing--Australia. I. McManus, Phil,
1966- II. Title. III. Series.

  HB2135.C66 2011
  307.2'60994--dc22

                                                                                    2010048853

ISBN 978-0-7546-7511-2 (hbk)
ISBN 978-1-138-26016-0 (pbk)

Contents

List of Figures		*vii*
List of Tables		*ix*
Acknowledgements		*xi*
Preface		*xiii*
Abbreviations		*xvii*

1	Rural Revival?	1
2	Leaving the City	23
3	Country Week	39
4	Strategies: 'In It to Win It'	63
5	A Place on the Map?	85
6	Going to the Show	97
7	Taking to the Country	117
8	The Good Resident	137
9	Living the Dream? A Retrospective	169

Bibliography	*179*
Index	*189*

List of Figures

1.1	New South Wales places	xviii
1.2	Queensland places	xviii
1.3	Population change in New South Wales and Queensland 2001-6	7
3.1	Passport for NSW Country Week Expo 2009	45
3.2	Oberon stall Country Week Expo 2010	47
3.3	Parkes and Elvis Country Week Expo 2010	47
3.4	Torres Strait dancers Country Week 2008	48
3.5	Gunnedah's clock display Country Week 2008	49
3.6	NSW and Queensland Councils. Country Week 2007	60
4.1	Oberon in place, Country Week 2008	67
4.2	Employment Board, NSW Country Week 2010	75
4.3	Putting Grenfell on the map, Country Week 2008	80
6.1	Expressed influences on migration, Country Week 2006	99
7.1	Age profile of newcomers to Oberon and Glen Innes	119
8.1	Country Week Expo brochures	138

List of Tables

1.1   Resident populations of selected NSW urban centres 1976-2006   8
1.2   Resident populations of selected Queensland urban centres 1976-2006   9
1.3   Estimated resident population, selected urban centres and local
       government areas, New South Wales, 1976-2006   11
3.1   Exhibitors at the 2008 NSW Country Week Expo   51
4.1   Themes present in selected rural newspapers available at
       Country Week Expo   70
6.1   Preferred places for relocation, NSW, 2006   100
6.2   The most frequently sought-after places, 2007   104
7.1   Reasons for moving to Glen Innes and Oberon   121

Acknowledgements

This book could not even have been contemplated without the genial and enthusiastic support in so many ways of Peter Bailey, the CEO of Country Week (now the Foundation for Regional Development Limited). He took us under his wing when we attended our first Country Week Expo in 2006, supported our students, Amanda Tsioutis and Marita Cuomo, when they undertook subsequent theses on various facets of Country Week, and stimulated our various efforts at analysing the strategies and successes of Country Week. We also owe his wife Jenny grateful thanks for her similar constant support.

It goes without saying, but it cannot be allowed to do, that we could not have done very much without the assistance of the various councils and other groups who attended the Expos and were always happy to make their views, ideas and documents available to us. Over the years we came to know some of them well, but perhaps none more so than the ever innovative team from Moree and the enthusiastic crew who always brought Elvis out from an active retirement in Parkes.

Equally we, and our students, would not have got very far without the willingness of visitors to the Expo to fill in questionnaires, and submit to multiple questions, and the willingness of recent residents in Oberon and Glen Innes to answer questions. Inevitably this also means that we are extremely grateful to Amanda and Marita, and to Lionel Brown who led the way, for permission to use some of their work and insights in this book. We are also grateful to Nathan Wales and Andrew Wilson for assistance with the maps.

Phil would also like to acknowledge the support of Jennifer Barrett and Caitlin McManus-Barrett – researching and writing a book means time away from other roles and activities. John apologises to the Queens Park football team for not being on that particular paddock more often. They may not have noticed. Hopefully this book shows that it was time well spent. Speaking of time, we would like to thank the patient people at Ashgate for commissioning, supporting and waiting hopefully for this book.

Unless otherwise stated quotations are from visitors to Country Week Expos, council officials at the Expos or from leaflets, local newspapers and other council promotional material, as is relevant. Most have been given appropriate anonymity.

John Connell and Phil McManus
School of Geosciences,
University of Sydney.
2011

Preface

Is it possible to re-populate and otherwise support declining rural and regional areas? If so how might this best be done? This is a crucial and complex issue for many countries that have experienced both rural decline and metropolitan congestion. This book examines the problems of regional development in Australia, centred on population decline and stagnation, and particular efforts to achieve a population turnaround that might 're-populate' the inland: referred to by some as a 'tree change'. Re-populating rural areas to secure rural revival is an issue that has become important in many developed countries. The response in Australia, through the concerted and partly combined efforts of concerned councils, enables critical appraisal of strategies that challenge rural decline and urban migration.

In a broader context ideas of rurality and urban–rural migration are being questioned in a number of countries and theoretical contexts. This book extends and links issues of rurality, counter-urbanisation, contested notions of rural gentrification and lifestyle migration, and rural place marketing and place branding, to explore an innovative, organised place-marketing activity that simultaneously involves cooperation and competition among participants. This activity, the annual Country Week Expo, became known as the Country and Regional Living Expo late in 2009, but here we retain the name Country Week (CW), its name at its inception in 2004 and while most of our research was undertaken. As the Country Week Chairman, Anthony Fox, noted in his opening of the 2009 Sydney Expo, these events are 'unique in Australia' and he 'had not heard of them taking place anywhere else in the world'.

This book focuses primarily on New South Wales (NSW) (see Figure 1.1), and to a lesser extent Queensland (see Figure 1.2), because NSW is a state where institutional efforts at decentralisation have been longer established, where there is more secondary literature, and because of our own greater familiarity with the state. The secondary focus is on the efforts of one particular private sector-run organisation, Country Week (CW), to stimulate such a population turnaround. Country Week is a rare example of collective place-marketing, focused directly on households, rather than on promoting economic activities or regional development grants and infrastructure. In practice CW is a three-day city Expo which has operated since 2004 in NSW and subsequently in Queensland. It involves rural and regional councils, and key government departments, publicising the advantages of rural and small town residence and encouraging households to relocate from cities to places where population growth is weak or non-existent, and/or where employment opportunities are available.

At the core of Country Week there is a crucial paradox: if regional centres are the positive places that CW publicity emphasises they are, why then does CW exist to encourage people to move to such places, and why is so much advertisement necessary? This book seeks to resolve this paradox, and examine the role of individual households in rural revitalisation. It looks in particular at the ideology of CW and highlights the perspectives of participating councils, and the viewpoints of visitors to the Expos who may, perhaps, move from the city to the country.

In Australia recent urban–rural migration has often been characterised as 'sea change' (or 'tree change', where movement was inland), mainly involving retired or semi-retired households making personal decisions to move from cities to pleasant coastal areas for what have been said to be lifestyle reasons. By contrast Country Week was primarily designed to attract younger entrants into the rural workforce, especially tradespeople and other skilled workers, to inland areas, some of which were declining. Subsequently, in Queensland, it sought to encourage workers for regional economies riding a mineral resource boom. In either case, despite partially trading on images of rural idyll (countryside, community, space, free time, security etc), the outcome was intended to be quite different from earlier processes of rural gentrification and counter-urbanisation. This raises questions about rural identity, the understanding of socio-economic status in rural migration, the role of strategic marketing and the significance of scale.

Population turnaround, 'tree change' and Country Week are all situated within diverse discourses of rural history, regional policy and changing migration patterns. This analysis thus seeks to link behavioural perspectives (processes of migration at the household level and decision-making at the local government level) with an understanding of macro-scale processes of socio-economic restructuring and changing political agendas. The Country Week Expos offered an exciting opportunity to forge these links, as CW has both emerged from early discourses and the 'realities' of rural decline, and in turn contributed to them though being a conduit between micro-scale decision-making and macro-scale restructuring. The phenomenon of 'marketing country' has not been documented elsewhere, despite rural regeneration being a critical issue in many western countries. Place marketing has been undertaken by local authorities in Sweden, but without centralised place-marketing fairs such as Country Week Expo, hence competition between councils was more evident than collaboration (Niedomysl 2004, 2007). What has occurred in Australia may thus be of distinct interest elsewhere.

Research Methodology

The primary research for this book involved both quantitative and qualitative research including questionnaire surveys, in-depth interviews and general observations and discussions, at the 2006, 2007, 2008, 2009 and 2010 Country Week Expos in Sydney, NSW, and the 2007 and 2008 Country Week Expos in

Brisbane, Queensland. The 2006 and 2007 questionnaires elicited the reasons for people's attendance at Country Week, sought their perceptions of country towns and regions, and examined how the Country Week Expos had influenced their perceptions. The questionnaire was administered over the three days of the Expos as exit surveys, so that the responses reflected what visitors had just experienced, rather than what they might hope to achieve. In 2006 in Sydney some 576 responses were received. These were supplemented by similar more detailed questionnaire surveys of randomly sampled participants both in Sydney and Brisbane in 2007. A series of semi-structured interviews was also conducted from 2006 to 2010 in Sydney, usually at least 45 minutes long, with local government representatives from participating councils. At various times, at the various Expos, these were supplemented by extensive discussion with the participants, whether visitors or stall holders, thus giving some measure of the evolution of visitor and council strategies.

Before looking in depth at Country Week and contemporary urban–rural migration in Australia, it is necessary to examine why some form of rural revival might be needed. The following chapter traces the history of discourses about the problematic countryside, within the diversity of rural and regional Australia, and the contrasting fortunes of different locales, but especially those where rural decline has been pronounced. It therefore considers the notion of 'country-mindedness', and broader questions about 'rurality'.

Abbreviations

CW	Country Week
CWA	Country Women's Association
GDP	Gross Domestic Product
LTD	*Live the Dream*
NSW	New South Wales
SA	South Australia
WA	Western Australia

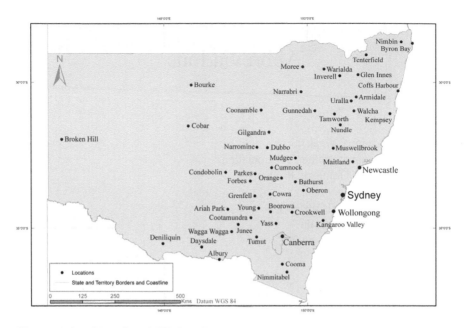

Figure 1.1   New South Wales places

Figure 1.2   Queensland places

Chapter 1

Rural Revival?

Much rural policy is influenced by perceptions of rural *disadvantage*. Rural
communities do face special problems and challenges ... [but] it is important to
note the positive aspects of life outside the metropolis. For most people, living in
these areas is a positive rather than a residual choice.

(NSW Government 1995: 8-9; emphasis in original)

In recent decades most rural areas, unless blessed with valuable natural resources or
tourism potential, have struggled. Agriculture has declined, providing fewer jobs,
and rural infrastructure and services have often failed to keep pace with those in
larger urban areas. Virtually throughout Western Europe, North America, Australia
and New Zealand similar trends have been apparent. As rural areas have moved
closer to a post-agricultural future, and towards a so-called post-productivist
countryside, once simply a place of 'rural dilution' (Smailes 2002), populations
have tended to decline and age. The exceptions are towns within commuting
range of large cities, or in particularly favoured regions, especially coastal areas
of second homes and retirement. By contrast in more distant areas the viability
and sustainability of rural and small town life has been questioned. This has raised
concerns about the future of particular places, and even entire regions.

It has been rightly claimed that 'Australia has become a land where regions
matter' (Beer et al. 2003: 1), but real variations exist within Australia over their
importance. Queensland, for example, with several large cities, is more regionally-
oriented than NSW, where NSW is sometimes seen as merely an acronym for
the largest coastal cities of 'Newcastle-Sydney-Wollongong', separated by a
'sandstone curtain' from the bush. In Australia, regions have often become
identified with the emerging concept of a somewhat disadvantaged 'rural and
regional Australia' (Pritchard and McManus 2000, McManus 2005). The existence
of significant, and growing, variations in regional prosperity has prompted calls
for Federal Government involvement in regional development, on the basis that
it has the financial capacity, due to its revenue generation, to overcome state and
regional conflicts (Maude 2003, Beer 2006). The Federal Government, under the
Liberal Party Prime Minister, John Howard, which governed for 11 years until
2007, shied away from involvement. State governments operated different forms
of regional development, with South Australia (SA), Western Australia (WA)
and NSW 'essentially funding regional development boards outside the capital
cities only' (Beer 2006: 122). Different structures and practices influence regional
development in various states, while the existence of development bodies does
not necessarily equate with development, employment and rural revival. The
role of local government is significant, but most councils perceive themselves as

hamstrung by limited funding and technical capacity. In a survey of 302 local governments throughout Australia, two key areas that they saw as 'least effective' were 'inward investment, promotion of region' and 'training, skills, and supply side labour market interventions' (Maude 2003: 124). At both regional and local level structural development has been weak.

The latter decades of the twentieth century saw an increase in the number of narratives of rural decline in Australia (Country Shire Councils Association and Country Urban Councils Association Working Party 1990, Taskforce on Regional Development 1993, NSW Government 1995) and their parallel emergence in countries like Canada and the USA. In Australia these narratives were mobilised and sometimes contested, either politically, through the rise of Pauline Hanson and the One Nation Party in the 1990s (McManus and Pritchard 2000) or by academic analysis of changes in rural and regional Australia. In the twenty-first century, the diversity of what constitutes 'rural and regional' or 'the country' is recognised by many people, but is entwined with perceptions of rural decline and images of drought and accompanying concerns about climate change and water availability. This book explores and analyses one event, Country Week (CW), designed to address perceptions of both rural decline and regional promise in NSW and Queensland, and recent population mobility in regional Australia. In order to do that it is first necessary to examine rural history.

Why do discourses of rural disadvantage hold sway in countries such as Australia? How did the countryside or, in Australian lexicon, 'the bush', come to be constructed as problematic? This chapter examines the origins and development of discourses about the problematic countryside, the 'rural' and the 'bush', and their link to rural and regional development. It considers the relevance of these discourses, and points to the diversity within rural and regional areas and the contrasting fortunes of different locales. It further considers the notion of 'countrymindedness', which argues that only 'real' goods produce wealth, and that rural industries such as agriculture not only produce wealth but also produce 'good people', unlike the situation in large cities. Is this seemingly antiquated notion of countrymindedness dead, or does it appear in new guises in the early twenty-first century? This book suggests that something akin to countrymindedness does still exist, if without the widespread influence of previous eras. It is certainly a central element in discourses surrounding the three day CW Expo in Sydney and Brisbane that seeks to revive rural and regional Australia through place marketing of rural and regional locations with the intention of encouraging urban–rural migration.

In order to understand recent discourses of rural decline and of rural and regional development, it is necessary to situate 'the rural' in the history of developed countries. While the specific history of each country is unique, there are clear parallels with the USA, Canada and New Zealand in relation to technological change, agricultural employment and the socio-economic development of rural regions. Australia's rural history is a necessary starting point in developing an understanding of the uniqueness and significance of CW.

A Brief History of Rural Australia

For clarity, the history of rural Australia is separated into three time periods. Though these periods are not arbitrary, they are neither a definitive division of time and nor do they necessarily reflect the most significant events in regional Australia, or in particular industries in rural Australia. They also relate to different phases of migration, with immigration booming until the depression of the 1890s, and post-World War Two migration increasing the population of state capital cities and some parts of rural and regional Australia.

Rural Australia pre-1891

Indigenous people have lived in Australia for at least 40,000 years. Prior to the arrival of European explorers and settlers after 1788, they lived in their specific 'country', where they hunted, gathered, used the land and managed the water in order to survive. This was unrecognised by the European settlers and, without signing any treaties, Aborigines were dispossessed of their land so that it could be 'worked', based on English notions of agriculture, rural villages and the division of land into fenced fields. The application of these notions to Australian conditions not only meant dispossession, it also meant that European settlers struggled to survive in the Australian environment.

Agriculture was vital for the success of Australian colonies, because the distance and dangers of importing food from England meant that self-reliance in food sources was paramount. By the end of the first century of colonisation (1888) agriculture was well-established in most colonies, and was prospering and expanding westward in NSW particularly with the extensions of the railway network. However prosperity was based on overstocking during periods of favourable weather patterns: a practice that was to have severe economic, social and environmental implications when subsequent drought conditions dominated. As Jeans (1972: 13) observed: 'the 1890s proved to be particularly harsh years in which the optimism of previous decades was destroyed on both the pastoralist and farming frontiers'.

The colony of Queensland came into existence in 1859, and the 1860s and 1870s were crucial in establishing the settlement patterns of the colony as, during this time, infrastructure was built to support the pastoral, mining and agricultural industries. Despite the drought, economic depression and political upheavals associated with the Shearers' Strike in 1891, railway expansion continued westward to support pastoral activity (Fitzgerald et al. 2009). This infrastructure contributed to the decentralisation of the population: a legacy being that Brisbane has always had a low level of primacy compared with the other mainland Australian state capitals.

While rural Australia became associated with agriculture (both pastoralism and farming) because of its prosperity in the late nineteenth century, other primary industries, and quasi-agricultural industries, were present in various parts of the country. In NSW, for example, there were several fishing ports, a whaling industry in Eden, coal mining in Lithgow, the Illawarra and Hunter regions south and north of Sydney and further north around Gunnedah, thoroughbred breeding near Scone in the Upper Hunter region, while viticulture had been introduced into much of the Hunter Valley (although this collapsed when cheap imports from SA flooded the market in 1901, at the time of Federation. In northern NSW and in Queensland, sugar cane was a major industry. The ability of wealthier agricultural interests to bind 'the bush' with 'agriculture', rather than other industries, was significant because it laid the foundations for discourses of rural decline that were to permeate rural policy deliberations in the late twentieth century.

Rural Australia 1891-1945

The east coast of Australia had been booming from the 1850s onwards, when gold was first discovered in NSW, and in the early 1860s in Victoria (Davison 2005). This boom came to a sudden end in 1891 when a combination of money market problems in London, severe and prolonged drought and falling wool prices resulted in a fall of about 30 percent in the GDP of Australia (despite a gold mining boom in WA) in the period 1891-1895. Sheep numbers in NSW, which had more than trebled in the 20 years between 1870 and 1890, plummeted as the availability of food declined and wool prices dropped. In 1892, at their peak, there were nearly 58 million sheep in the state; by 1900 this figure had fallen to approximately 40 million and did not reach the earlier figure again until 1942 (Jeans 1972).

Until the twentieth century the pastoral industry had 'provided the bulk of the export trade of the colony and was the chief stimulus to regional development' (Jeans 1972: 10). It is not surprising that the demography of NSW reflected the needs of the pastoral industry, and the constraints of available technology such as rail, chilling and communications: most country towns were small places and Sydney was 19 times larger than the second largest place (the inland mining town of Broken Hill, with 27,000 people) and only Newcastle (14,238) and Goulburn (10,612) had more than 10,000 residents at the 1901 Census. By contrast, in 1911, unlike all other state capitals, Brisbane held less than a quarter of Queensland's population.

After the prosperity of the late nineteenth century, 'by the early 20th century, however, young people were drifting from the countryside to the city, and the inland towns had begun their long decline' (Davison 2005: 39). It was the first era of rural–urban migration in Australia. A period of rural re-population occurred after World War One with the initiation of Soldier Settlements, and the introduction of measures to improve the quality of life in rural areas. By 1920 a national Australian Country Party had been formed (after earlier state based prototypes beginning in WA in 1913) and by the 1930s improved roads and communication services as well as freight subsidies assisted people in rural areas (Davison 2005). This helped maintain population in rural areas, as did the Great Depression that resulted in much unemployment in cities and some people moving from the city to search for work. Yet, while populations were maintained, poverty and hardship existed in many places.

Rural Australia 1945-present

The conclusion to World War Two saw a marked increase in the productivity of Australian agriculture, largely following the replacement of human and animal labour by machines. Politicians from all parties had positive visions of rural development. In Queensland, the Australian Labor Party government, and a Premier 'imbued with the agrarian ideals of former premiers ... saw a decentralised Queensland based on farming, pastoralism and mining as a way of avoiding the social problems of overcrowded cities: there was also a barely articulated sense that work in the primary industries created virtuous, productive citizens' (Fitzgerald et al. 2009: 120). In NSW the success of irrigation schemes on the Murrumbidgee and Murray rivers offered a new vision splendid for agriculture. Increased productivity was accompanied by good wheat and wool prices, resulting in rural prosperity, but displacement of people by machines had a deleterious impact on rural areas, especially on small agriculturally-dependent communities. Small towns within commuting distance of a larger centre were often most severely affected as new transport technology enabled people to travel further for goods and services, and the rationale for many smaller service centres disappeared. Visions of rural growth were further undermined by the tendency of post-war migrants to settle in the state capitals, though mining towns continued to grow in inland Queensland. Over time population losses through migration, as younger people left the country to search for employment (Chapter 2), translated into an aging residual population and a gradual population decline as deaths exceeded births.

Where agriculture retained vitality production often changed. Sheep numbers in NSW have again declined to their lowest levels in a century, cattle production has fallen and many abattoirs closed in the last two decades, while the area of wheat cultivation has expanded. Salinity has affected some inland areas. However boutique wineries have increased in numbers and wine production has soared at least until the recent global financial crisis and gluts on overseas markets. There has been an increasing agricultural diversity, ranging from irrigated agriculture (including cotton and avocadoes in northern NSW and south-west Queensland, and rice growing in the Riverina), olives, gourmet food (including cheeses and organic products), and cool climate wines in Orange and Stanthorpe.

The 1960s and 1970s saw population decline in many parts of rural and regional Australia, notably in the inland agricultural areas, despite some local, state and national strategies to avert emerging economic and population imbalances (Country Shire Councils Association and Country Urban Councils Association Working Party 1990, Pritchard and McManus 2000). The most prominent of these strategies, initiated by the Labor Party Federal Government in the early 1970s, attempted to reduce population growth in the capital cities by encouraging relocation to selected inland sites, but the outcomes were limited and the initiative lapsed (Chapter 2). At least one inland city has grown quite rapidly: 50 years after it was established as the national capital, Canberra had become the largest of Australia's inland cities, reaching a population of 70,000 in 1963, with more rapid growth to follow.

As the few growth centres demonstrated, rural and regional Australia was neither homogeneous nor uniformly in decline. In NSW between 1981 and 1993, 11 of the 12 statistical regions increased in population; the only decline was experienced in the Far West Statistical Division, which showed population losses at every census count in the century. Growth was slight and small population losses in one or more census periods occurred in the Central West and Murrumbidgee (NSW Government 1995). In Queensland a similar pattern emerged with most statistical divisions experiencing population growth, especially on the coast, and only the Central West losing population between 2001 and 2006 (Figure 1.3). Yet despite slow population growth, within both of these inland regions, some particular towns still prospered.

The existence of prospering and growing towns surrounded by smaller centres experiencing population decline suggested the idea of 'sponge cities': rather larger regional centres with superior infrastructure, where growth occurred, in distinct contrast to smaller nearby settlements. Services, especially government offices, and consequently regional populations, have become increasingly concentrated in the former – large towns such as Orange, Wagga Wagga and Tamworth in NSW and Toowoomba and Rockhampton in Queensland – at the expense of the latter. With greater access to cars, and thus long distance commuting and shopping, differential structures of growth and regional centralisation of services were at the expense of the continued decline of local services in small towns. Burnley and Murphy (2004) thus argued that population growth in inland regional Australia was a result of economic growth in larger centres, where towns such as Tamworth and Orange effectively sucked the life out of smaller towns and the surrounding countryside. Moreover that process had been going on since the rise of car ownership half a century earlier. The towns most at risk of losing population were those located in the catchment areas of the sponges (Maher and Stimson 1994) rather than more remote settlements. Thus Cooma, with 10,000 people, was big enough for 'chain stores such as Coles and Woolworths and *critical attractions* such as McDonald's and KFC' drawing business from Nimmitabel, 40 kilometres away (Higgins 2009: 9; our italics). Towns like Nimmitabel, with just 240 people, are close to becoming ghost towns, with empty houses and few prospects of regeneration (Forth and Howell 2002). Cooma, and similar sponges, drew in shoppers from such dwindling towns, but were too distant for commuting from them.

Continuing the hydrological metaphor, the sponge city migration thesis may not however hold water. Migration trends over the period 1986-2001 in Dubbo and Tamworth, two NSW towns often cited as 'sponge cities', were more complex and the population 'soaked up' from surrounding settlements was a relatively minor part of their growth. Such larger towns, both with populations over 35,000, primarily drew new migrants from beyond their immediate hinterlands, mainly from Sydney

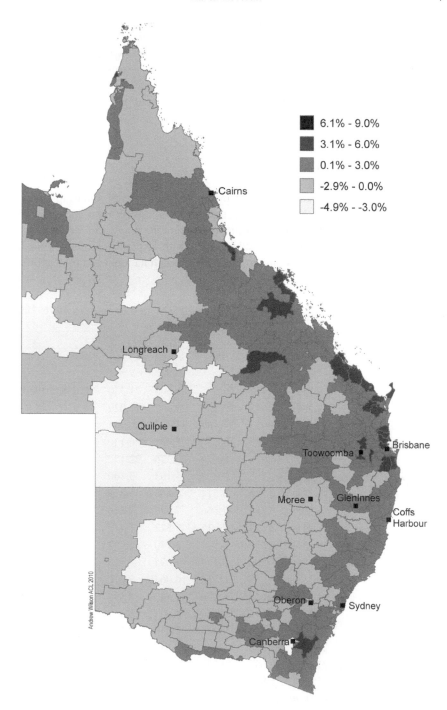

Figure 1.3   Population change in New South Wales and Queensland 2001-6

and outside the state (Argent et al. 2008). The 'sponge city' phenomenon is at best only partially true, though it is incontrovertible that larger towns are playing some part in sucking the life out of smaller towns, and that smaller towns have continued to decline fastest as their residents moved on or simply died. It is useful therefore to explore in more detail population changes in regional areas, situating recent Australian trends in an international context.

Population Decline in Rural Areas

Population decline is consistently most pronounced in smaller inland towns and rural areas (Taskforce on Regional Development 1993, NSW Government 1995, Hugo 2005, Murphy 2006). For Australia as a whole the population in rural areas (defined as having less than 1,000 residents) fell from 17 percent of the population in 1966 to 14 percent of the population in 2001, though national and metropolitan

Table 1.1   Resident populations of selected NSW urban centres 1976-2006

Urban Centre	Census Years						
	1976	1981	1986	1991	1996	2001	2006
Armidale	19,711	18,922	19,525	21,605	21,330	20,068	19,485
Bathurst	18,589	19,640	22,237	24,642	26,029	26,920	28,992
Broken Hill	27,647	26,913	24,460	23,263	20,963	19,753	18,854
Byron Bay	2,525	3,187	3,730	5,001	6,130	5,919	4,981
Coffs Harbour	12,197	16,020	18,074	20,326	22,177	25,828	26,353
Condobolin	3,273	3,355	3,229	3,163	3,100	3,054	2,847
Cootamundra	6,384	6,540	6,314	6,816	5,879	5,486	5,566
Deniliquin	6,865	7,354	7,566	7,895	7,816	7,786	7,431
Forbes	7,754	8,029	7,915	7,522	7,445	7,094	6,954
Glen Innes	5,953	6,052	5,971	6,140	6,092	5,706	5,944
Gunnedah	8,689	8,909	9,144	8,874	8,315	7,871	7,542
Moree	9,359	10,459	10,215	10,062	9,270	9,273	8,083
Mudgee	5,724	6,015	6,576	7,447	8,195	8,619	8,249
Narrabri	6,951	7,296	7,246	6,694	6,417	6,235	6,102
Oberon	1,982	1,937	1,840	2,236	2,545	2,516	2,473
Orange	26,254	27,626	28,935	29,635	30,705	31,923	31,544
Tamworth	27,273	29,657	30,729	31,716	31,865	32,440	33,473
Tumut	5,569	5,816	6,099	5,955	5,915	6,243	5,925

Note: Coastal locations are shown in italics.

Source: ABS Census data.

populations grew steadily during that time. Population decline was most concentrated in the dry farming areas of most states, pastoral areas, and in remote mining communities, notably Broken Hill (Hugo 2005). By contrast coastal areas, especially in NSW and Queensland, grew rapidly, largely as a result of the 'sea-change' migration from the larger cities. Hugo (2005: 78) noted that 'a significant amount of growth is occurring in towns at and just beyond the commuting limits surrounding major cities, so that some would argue there is not so much a counter-urbanisation trend that is occurring but a new, more diffuse form of urbanisation' (see Chapter 2). Population growth, or 'turnaround', in coastal Australia, had few parallels in inland regions, other than in a few mining areas in Queensland and WA, as counter-urbanisation clung to the coast. A profound demographic shift was occurring.

Numerous inland towns have stagnated or declined over many years, a trend continuing in NSW and evident in Queensland and other Australian states, despite considerable diversity (see Table 1.1 and Table 1.2). In both states coastal towns have grown relatively rapidly, small inland centres have declined and larger regional centres, such as Toowoomba, Bathurst and Tamworth have grown. Since

Table 1.2   Resident populations of selected Queensland urban centres 1976-2006

Urban Centre	Census Years						
	1976	1981	1986	1991	1996	2001	2006
Barcaldine	1,443	1,432	1,427	1,530	1,592	1,497	1,337
Cairns	39,305	48,557	54,862	64,463	92,273	98,981	114,762
Charleville	3,802	3,523	3,588	3,513	3,327	3,506	3,278
Cunnamulla	1,897	1,627	1,697	1,683	1,461	1,356	1,217
Dalby	8,997	8,784	8,338	9,385	9,517	9,731	9,778
Emerald	3,161	4,628	5,982	6,557	9,345	10,092	10,999
Gold Coast	94,014	135,437	163,332	225,773	274,157	376,573	402,648
Hughenden	1,811	1,657	1,791	1,592	1,444	1,395	1,154
Longreach	3,354	2,971	3,159	3,607	3,766	3,648	2,976
Moranbah	4,053	4,362	6,883	6,525	6,508	6,133	7,133
Quilpie	728	694	780	624	730	645	560
Roma	5,898	5,706	6,069	5,669	5,744	5,987	5,983
St. George	2,095	2,204	2,323	2,512	2,463	2,779	2,410
Thargomindah	n/a	n/a	259	267	215	217	203
Toowoomba	63,956	63,401	71,362	75,990	83,855	89,338	95,265
Warwick	9,169	8,853	9,435	10,393	10,947	12,011	12,562

Note: Coastal locations are shown in italics.

Sources: ABS Census data.

the 2006 Census, some towns that were declining for a number of years, such as Gunnedah, have experienced a resource boom, so trends do not necessarily equate with destiny. By way of contrast, some single industry towns were particularly hard hit by the global financial crisis. The mining town of Cobar, with around 6,000 people, lost 10 percent of its population (with 600 redundancies) in six months in 2008. All mining apprentices were laid off, outmigration ensued, many local service industries experienced massive declines in revenue, several closed and 'for sale' notices appeared on many businesses and homes: a microcosm of a downturn (Jensen 2009). Nonetheless underpinning the presence of many councils and other organisations at the CW Expos is the desire to draw more people to regional areas to boost their populations, and take up employment and business opportunities, whether or not they have experienced decline or wish to build on existing growth.

Throughout Australia rural areas, characterised by farming and related services in small towns, have been the main areas of population loss, whereas larger towns have retained populations or increased in size, unless they are very remote and not experiencing a resource boom. Rural decline has been hastened by increased farm efficiency, farm amalgamations, continued mechanisation and more capital-intensive production alongside growing disdain for agricultural employment. The end of the long economic boom emphasised this downturn. The ongoing population decline that is evident inland is most pronounced in the rural areas, that is in the local government areas surrounding the towns (Table 1.3), and has sometimes resulted in the abandonment of both farms and farmhouses. Structural changes to some local government areas preclude more detailed analysis but the distinction between the town and country populations has intensified. Even tiny Oberon is growing relative to nearby villages. Rural towns are nonetheless 'hanging on', despite the greater loss of population in rural areas, though there is a diversity of demographic scenarios in rural and regional Australia, and even seemingly similar mining towns such as Cobar and Gunnedah have experienced marked changes in fortune over time. The differences can result from various combinations of factors, ranging from local initiative and acumen through to corporate decisions made in distant boardrooms; from the experience of a disaster, such as drought or flood, through to commodity price fluctuations on international exchanges, or taxation policies that may make particular locations suitable for particular kinds of mining or agriculture. There is no single determinant of success or failure.

The decline in population tends to mask two other trends within inland rural settlements; the aging of the population *in situ* and the absolute and proportional increase in indigenous populations in most towns. Both are significant for the economic health of towns, because neither are regarded as high-powered consumers. This has obvious effects on the viability of businesses in rural towns, with follow-on impacts in the loss of skilled people, other community services, and so on.

Rural decline in NSW is no isolated phenomenon. Despite the exceptional population growth of Queensland, from 2.4 million people in 1982 to 4.3 million

Table 1.3   Estimated resident population, selected urban centres and local government areas, New South Wales, 1976-2006

Town and Local Government Area	1976 Pop.	1981	1986	1991	1996	2001	2006	Pop change 1976-2006	% change 1976-2006
Condobolin	3,273	3,355	3,229	3,163	3,100	3,054	2,847	-426	-13.0
Lachlan Shire	8,583	8,403	8,049	7,694	7,588	7,180	6,669	-1,914	-22.3
Glen Innes	5,953	6,052	5,971	6,140	6,092	5,706	5,944	-9	.2
Severn Shire	3,053	3,049	3,093	3,138	2,915	2,789	2,836	-217	-7.1
Forbes	7,754	8,029	7,915	7,522	7,445	7,094	6,954	-800	-10.3
Forbes Shire	10,934	10,993	10,736	10,343	10,370	9,691	9,316	-1,618	-14.8
Moree	9,359	10,459	10,215	10,062	9,270	9,273	8,083	-1,276	-13.6
Moree Plains Shire	15,411	17,229	17,018	16,918	15,364	15,680	13,976	-1,435	-9.3
Oberon	1,982	1,937	1,840	2,236	2,545	2,516	2,473	491	24.8
Oberon Shire	3,835	3,845	3,845	4,258	4,608	4,817	5,030	1,195	31.2
Tumut	5,569	5,816	6,099	5,955	5,915	6,243	5,925	356	6.4
Tumut Shire	10,987	11,399	11,507	11,175	11,398	11,172	10,801	-186	-1.7

Source: ABS Census data.

people in 2008, parts of the state experienced decline. While the South-East Queensland region around Brisbane-Gold Coast-Sunshine Coast has attracted population from other locations, mainly from southern states, and coastal locations have generally experienced significant population growth, inland regions have scarcely increased their populations even during a mining boom. In mining boom centres (such as Bowen and Mt. Isa) populations also age and such towns have sought to draw workers from elsewhere, incidentally reducing the chances of regions without well-paid mining employment, such as Rockhampton, from gaining new workers. Similar trends have been even starker in WA, while losses of workers from various states to the mining boom in WA and Queensland has had considerable flow-on effects on other local communities. Wheatbelt towns in WA not only lose young people who migrate to the city or to mining boom-towns, they are failing to attract them back after a period of city life, and encourage others to move, mainly because of negative perceptions of rural social life (Davies 2008). Inland towns in Queensland, where mining is absent, have lost population in recent years, including Charleville (3,626 to 3,513), Quilpie (686 to 589) and Longreach (3,353 to 3,124) between 2001 and 2009 (Queensland Government 2010). Each of these towns is too far from a growing city to experience any possible spill-over effects from urban growth.

While the mining boom has given a certain distinctiveness to WA and Queensland, population shifts in NSW and Queensland are otherwise broadly similar to elsewhere in Australia, where sea change, counter-urbanisation and regional decline are widespread but mainly on a smaller scale. Tasmania too exhibits similar trends, but the southern states have also been characterised by migration to warmer northern climes and, until quite recently, Tasmania has been the only state to experience population decline.

Whither the World?

Broadly similar population shifts have characterised rural areas in much of the developed world, though few have been well documented. In the relatively thinly populated inland states and provinces of USA and Canada, population decline continues. In Kansas, for example, many western counties experienced population decline between 2001 and 2008. The rates of decline were astonishing: many counties lost about 10-11 percent of their population between mid-2000 and mid-2008, and this was simply part of a longer and wider trend (Wood 2008). In neighbouring Nebraska the story has been similar with population losses ranging up to 13 percent. Here too agricultural communities are remote from larger cities. Such severe population declines in the mid-western states have threatened the viability of various settlements: an inland decline that offers close parallels with Australia.

Canada too experienced similar trends. The entire province of Saskatchewan lost population in the 20-year period 1986-2006, and many towns showed similarities with Australia in that their population decline was ongoing, and there was little opportunity for revival through overspill from a larger centre. Similar trends occurred in neighbouring Manitoba, where several small towns had striking losses, some greater than 10 percent over just the 2001 to 2006 period. Again such towns could mainly be characterised by remoteness from metropolitan centres. On a much smaller scale agricultural regions in New Zealand have lost population for similar reasons, as agriculture loses its viability and coastal cities offer greater attractions. Population decline in the predominantly rural region of Southland has occurred for some decades. Between 1976 and 1991 its population declined from 108,860 to 100,431, with a 'greater decrease among younger people, especially those aged 20-24' and the most rapid decline in small centres with fewer than 1000 people between 1986 and 1991 (Ministry of Agriculture and Forestry 1994). The rate of decline slowed in this century with the rural population rising in 'rural areas with high urban influence' and declining in the 'highly rural/remote area' (Statistics New Zealand 2008).

In more densely settled and smaller European countries parallel examples of rapid recent population decline are harder to find, and it is in the peripheries, such as the Celtic borders of Great Britain and northern Scandinavia, some of which are culturally distinct, where population declines have been significant, but usually more steadily and over a longer time period. All are countries where governments have tried a range of policies – centred on subsidised regional development strategies and incentives, salary supports and tax concessions and infrastructure provision – to halt and reverse declines

Population decline in rural areas is far from unique to Australia, and much of the rationale, the structures and their socio-economic consequences are similar. Population decline tends to affect most rural communities through the departure of young people, particularly girls and young women, in search of the three Es – education, employment and excitement – in big-city lifestyles. This can be part of a downward economic spiral that is as hard to reverse in NSW as it has been in Nebraska, Manitoba, Southland or the Orkneys.

Lost Youth

Unsurprisingly the combination of declining populations and services, aging populations and workforces, and even the loss of identity, in the face of global and national growth has been unwelcome for multiple reasons, not least because it seems the thin end of the wedge for an inexorable continued decline. Rural areas experience the loss of youth, often the real and potential 'best and the brightest', and face particular challenges to attract and retain skilled workers, such as doctors and nurses, and consequently struggle to retain key services.

Selective loss of youth is universal. In the small NSW town of Oberon, only three out of a sample of 23 school leavers over the period 1991-2004 had not moved away, though another three had left and returned (Brown 2006). In 2006 half the graduates of one Gunnedah high school had left the town within a year (Hall 2006). In the much smaller and more remote town of Ariah Park, NSW, only three out of 19 school leavers in 1999 remained there six years later, and one of them was about to leave (Lewis 2005). Full-time job opportunities for the 15-24 age group, and particularly for women, declined from 1986 onwards in a typical group of eight local government areas in regional NSW (Alston 2004) while those jobs that remain are relatively unattractive, with low wages and poor career prospects. For such small towns as Ariah Park, with a population of around 400, the future is bleak, hence leaving such places is 'a simple rite of passage of rural life – a way to get ahead and the way it's been for a long time' (Lewis 2005: 21). As one young person in rural NSW stated 'We are brought up with the fact that you are going to move out as soon as possible. Because you are going to get a job somewhere and it's not going to happen in X' while another noted reluctantly 'This is my home. I love living here. [But] you don't want to be a checkout chick all your life' (quoted in Alston 2004: 303). Getting away has become a youthful rite of passage, as part of a wider culture of migration, where migration is normative and success lies elsewhere (Easthope and Gabriel 2008). Youth migration continues to increase especially from areas that lack intensive agriculture or tourism development.

Young women have long been particularly likely to migrate from the smallest towns. This situation is ingrained, hence the popularity since the mid-2000s of the Australian 'reality' television series, *The Farmer Needs a Wife*, focusing on the needs of men who remained behind, and *Desperately Seeking Sheila* (which claims to 'take a warm-hearted look at a little known crisis affecting our blokes in the bush: an outback love drought'). Women who remained were less likely to be able to escape stereotypical female roles; in Glen Innes, in the 1980s at least, women even tended to blame themselves for not getting a good job so that 'attachment to locality [was] associated with rather low levels of self-esteem' (James 1989: 70). Men have more opportunities and are more at ease in a macho sporting culture that provides a sense of identity and belonging (Dempsey 1989, Easthope and Gabriel 2008). Employment is not the sole attraction of larger places. As one migrant from Ariah Park to Wagga Wagga observed: 'I would never go back to a small town ... I enjoy doing my own thing and not having everybody know you' (quoted in Lewis 2005: 21). Such patterns are loosely replicated throughout much of regional Australia as the future life blood of population growth is drained away, and small town populations age without replenishment.

A more detailed picture of youth migration since 1991 is available for Oberon, where the population aged under 29 fell from 1,450 in 1996 to 1,230 in 2006. Those who had left Oberon primarily sought further education (52 percent) or 'improved employment opportunities' (65 percent), though both reasons were entwined. Consequently most left town at the same time that they left high school; there was no reason to linger. Virtually no significance was attached to any other reasons; three people (out of 23) were influenced by a partner, and no more than two spoke of 'a desire to travel', 'family reasons', 'a change in lifestyle' or the wish to move from 'a small town' (Brown 2006: 40-42). Employment and education, as elsewhere, were critical. Of those who had left Oberon, more than half had jobs that were locally almost impossible: production manager for a fashion label, senior systems analyst, financial planner, radiographer, Aboriginal education officer and so on.

Not all who left Oberon went to Sydney though it did account for the largest number (nine out of 23). Five had moved outside NSW, to Queensland, Victoria or the Australian Capital Territory (Canberra), but the second largest group (five people) were in Bathurst, little more than 50 kilometres from Oberon, but with a much larger range of employment options and a small university (Charles Sturt University). Migration from Oberon was similar to that on a larger scale in NSW, where it was regional rather than necessarily to capital cities such as Brisbane or Sydney (Argent and Walmsley 2008) though initial moves to larger regional centres may merely be precursors to later moves to much larger cities.

Moving away offered different lifestyles. Some of those who had left Oberon for whatever reason were conscious of the limitations of small town social life, arguing that boredom led teenagers to experiment with under-age drinking, drugs and sex, while for younger adults social life largely revolved around drinking in pubs. Those who stayed often married young or had early pregnancies, situations the leavers sought to avoid (Brown 2006: 43). Correspondingly, as younger people left, some local recreational opportunities declined; loss of youth often meant the loss of numbers for sporting teams, real bastions of rural life, and a decline in the overall quality of life. Few found it difficult to leave, and most were happy they had gone, suggesting that return migration was not imminent, despite protestations of allegiance to Oberon, where all but one had close relatives. However, partly because so many had left, friends, including friends from Oberon, were more likely to be elsewhere. When those who had left were asked if they were likely to return to Oberon at some point, barely a third could even envisage this, while the rest thought that it was highly unlikely. The longer they had been away from Oberon, the less likely was return; more recent leavers who had yet to establish themselves elsewhere at least considered the possibility of return. But even those who thought they might return pushed that possibility into the distance, at least for another decade (Brown 2006: 49-50).

Those who had stayed in Oberon were relatively few, compared with the leavers, and were here more likely to be female (in a place where agriculture was of limited relative importance), married and with children. Out of a sample of 14 stayers, none had gone to university, compared with two thirds of those who had left Oberon, though more than half had taken TAFE (Tertiary and Further Education) courses. Tertiary education thus emphasised the selectivity of outmigration. As one Oberon resident pointed out: 'I think there are enough jobs [here] for the ones who stay and don't want to move up the corporate ladder – there are not so many jobs for uni graduates but plenty for tradespeople' (quoted in Brown 2006: 61). Those who had stayed were employed mainly in retailing, in the timber mill or in forestry. They had stayed primarily for lifestyle reasons (59 percent) – the friendly people and community, and their own families – and because they could find employment (71 percent). Lifestyle considerations were rather more important for those aged over 30, reflecting younger people's preoccupation with employment.

Those who stayed articulated the same kinds of reasons that, in a different context, were the sort of values promoted by the councils at CW (Chapter 4). In Oberon 'you know most people and who they are', 'it's one big family', 'the community is great for kids' and the 'people are wonderful'. A handful emphasised safety: Oberon is a 'quiet town that is safe' especially for children, or, quite succinctly, 'it's home and my family is here' (quoted in Brown 2006: 59). Indeed every one of the 14 stayers had close family in Oberon; some claimed to have all their relatives there. At much the same time the local newspaper, the *Oberon Review*, had conducted its own short survey of five residents and, somewhat predictably, they too emphasised the friendly people; one pointed out 'the sense of community, the friendliness of the people, the willingness to get behind community projects and the fact that we have four seasons' (*Oberon Review*, 10 August 2006).

Is Oberon typical? Oberon is a relatively young local authority area, where the median age increased from 33 years in 1996 to 38 years in 2006 (Australian Bureau of Statistics 2007), and a greater range of employment opportunities was available than in smaller towns. Oberon is similar in median age to many Sydney local authorities, larger centres such as Newcastle and Wollongong, and thriving regional centres such as Albury and Armidale. However the rate of aging is similar to more rural shires such as Liverpool Plains where the median age rose from 38 to 43, and Upper Lachlan from 38 to 44 (Australian Bureau of Statistics 2007). In terms of migration Oberon is actually much like the nation as whole. When the migration of over 5,000 youths across Australia was traced between 1995 and 2004, some 26 percent of those who had been living in a non-metropolitan area (outside the six state capitals) had moved into a capital city, often for education. Although return migration occurred it was not nearly equal to the rate of migration, even though some 30 percent of those who had first moved to a capital city had left it again for a smaller place a decade later (Hillman and Rothman 2007). That some stayed in Oberon, and not everyone enjoys capital cities, offers some opportunity and some optimism for the future of regional areas.

The Costs of Decline

While the media have tended to focus on the 'great divide' between 'rural losers' and 'urban winners' not all areas of regional Australia are in decline, and not all urban areas are economically successful (McManus and Pritchard 2000). Areas that have experienced the most significant population declines are inland, remote, and dependent on agriculture, where towns are small and services already limited. The costs of decline are diverse. Schools may close and teachers are less likely to be recruited, while health services slowly disappear, and the morale of those who continue to provide them declines as they must fill the gaps. Replacements become more difficult to recruit.

In much of rural and regional Australia declining production has meant the closure of both public facilities, such as post offices, police stations and schools, railway lines and stations, and of private sector activities, most evident in bank closures (Argent and Rolley 2000) but also in the widespread loss of chain stores, supermarkets and more specialist stores such as butchers. Reductions in public transport provision compound problems of access. Such trends have been accentuated by changes in government policy that have favoured user pays principles, market driven service allocation and the selective withdrawal of both government and private sector services in the quest for economies of scale, and smaller government (Alston 2004). The loss of jobs in such areas has threatened a much wider component of the community where agricultural households are dependent on some off-farm wage income for survival and retailing depends on expenditure of wages.

Even without losses of services, the imbalance in service provision between small towns and large cities is ever more palpable and a source of discontent, not only to the young, though they are most likely to rail against 'boredom ... nothing to do! No shopping malls, movie theatres or clubs' and 'not much variety in shops – no clothing stores' but also to older people who perceive 'no specialist medical services' and 'not enough aged care facilities' (Oberon residents quoted in Brown 2006: 80). Such deficits inevitably direct small town residents to larger centres. In turn this has ramifications for community, including the closure and amalgamation of churches, the inability of football and cricket clubs to field teams, or even the grassing over of bowling greens and tennis courts, all of which go to the heart of identity and community in small towns (Tonts 2005). During two decades of episodic rural crisis the tiny township of Daysdale, in southern NSW, experienced the closure of the post office, general store and primary school, which at its peak had 85 children and three teachers, while the local Australian Rules football and cricket clubs both folded for want of players (Argent 2008: 256-7). The burden of retaining other services falls on older, fewer or more distant people.

The increasing burden of community responsibility may test goodwill and the fragility of social capital when communities lack the resources to provide the infrastructure required for increasing or even retaining development options (Alston 2004, Argent 2008). Aging is of particular concern. The aging of regional Australia has been enhanced by the greater longevity of retirees, and by the movement of some retirees into regional areas (Drysdale 1991), with intensified pressure on services. Lack of facilities for aged care, including care homes, may effectively drive older people into larger urban centres, as it has done from Oberon (Brown 2006).

With the loss of population, revenue and services, councils have become economically unviable in many rural areas, and unable to meet infrastructure and service needs. Many have consequently been amalgamated both in NSW and Queensland, sometimes forcefully, as local communities sought to retain and defend a sense of scale and ownership of service provision (Dollery et al. 2009). Much of this institutional reform has taken place under the influence of neo-liberal economic rationalist programmes that have no basis in coherent national or state policies for regional development (Pritchard and McManus 2000, Beer et al. 2003, Maginn and Rofe 2007). Simultaneously the political influence of regional Australia has declined with the weakening of the National Party (formerly the Country Party), that sought to protect regional interests, despite the rise of committed independent parliamentarians in several regional areas. Political attempts to promote rural and regional Australia have included the rise of the One Nation Party, which, despite its numerous and significant shortcomings, rode the wave of disaffection experienced by rural people and outer suburban working-class populations, both of whom felt alienated by the existing political parties (see McManus and Pritchard 2000). But the One Nation Party faded fast and had limited direct influence. A Shooters Party, with members in the NSW upper house, was even less influential.

At various times in the past different political parties and other interest groups have sought to stimulate rural and regional development, most evident in the growth centre policies of the 1970s and sustained efforts by the National Party and others to support rural production. Underpinning many policies and strategies have been a range of attitudes to regional Australia, and to rural life, from within and outside the regions. Some part of that has come from an ideological basis in 'countrymindedness', where the ethos of rural Australia is perceived as somehow different from city life, which has informed not just political approaches to rurality but also the attitudes of residents and various components of the media. Such perceptions and their role in developing responses to rural decline have been significant.

Rurality, Countrymindedness, and Revival

Rurality is basically defined as the state of being rural. While this may initially appear uncomplicated, complexities are attached to constructing 'rural'. This is usually done in relation to its supposed opposite, urban, but cities have environmental and food-producing characteristics, just as industrialisation and commercialisation have taken place in rural landscapes (Lockie 2000, McManus 2005). Nonetheless the idea of the rural being a positive antidote to the crowded industrial cities of the nineteenth century dates as far back as Roman villas being constructed outside cities, and has been perpetuated through notions of rural idyll. This construct of the rural as a positive place in relation to the 'evil' city is based on stereotypes of what the country is, who 'belongs' there, and how they have 'traditionally' lived. All stereotypes are questionable, and ignore the dystopian aspects of rural life, including isolation, poverty, restrictive social values, and so on. The complexities and diversity of 'rural' locations means that it is difficult to define what is 'rural' (Phillips 2004). Inability to succinctly define 'rural' has led to various debates linking issues of rurality with counter-urbanisation (Sant and Simons 1993, Champion 1998, Mitchell 2004), contested notions of rural gentrification and lifestyle migration (Milbourne 1997, Phillips 2002, Paquette and

Domon 2004, Benson 2009, Hoey 2009, Stockdale 2009), rural place marketing (Niedomysl 2004, 2007) and place branding (Storey 2004, Cassel 2007, Mayes 2008). These themes are explored in what follows.

Diverse geographical and sociological studies have also sought to divide the 'rural' into distinct sub-groups and areas (Smith 2007). Lowe and Ward (2009) have proposed a sevenfold 'new rural typology' combining social and spatial phenomena: 'dynamic commuter', 'settled commuter', 'dynamic rural', deep rural', 'retirement retreat', 'peripheral amenity' and 'transient rural'. Such a division of rural landscapes into dominant characteristics has some value in highlighting the potential for various rural locations to be prospering, stagnating or dying, but there can only be a continuum. Many of the places discussed later would probably be labelled 'deep rural', that is places that 'seem to lack sufficient symbolic resources to attract in those socio-economic classes that are underpinning the vibrancy of the "commuter" categories' (Lowe and Ward 2009: 1324). That remains to be seen but certainly developing symbolic resources, and creating new opportunities, is invaluable.

Rural and regional Australia is diverse in its composition and in its relative fortunes. What might this mean for rural policy, and for people living in various parts of the country? At one pole it could mean that rural locations which are not prospering should be allowed, even encouraged, to decline into obsolescence. Indeed, in parts of Australia where drought has been persistent, farmers in 2007 were offered packages as high as \$170,000 to leave the land 'with dignity' (*Sydney Morning Herald*, 29 September 2007). Necessarily this approach results in further decline in already economically-stressed locations, but the alternative may be extensive subsidies. Such a scenario is not new, and the emergence of the Australian Country Party in Australia in 1920 was largely an attempt to ensure that the 'balance' did not swing so far away from rural Australia. The Country Party sought to consolidate rural power and redress what was perceived as a deteriorating rural–urban power relationship, with concomitant material consequences. According to Aitkin (1985), it was built on the notion of countrymindedness, a concept with similarities to the 'agrarianism' that flourished in the USA in the late nineteenth and early twentieth centuries (see Aitkin 1972, Kapferer 1991, Share 1995). In the history of Australia, the idea of countrymindedness has been widespread and influential. Its main tenets held shades of environmental determinism: Australia depends on primary production because only those who produce material goods add to a country's wealth. All Australians should therefore support rural industries, which are virtuous, cooperative and produce good people, in contrast to city life which is competitive and parasitical; city people are similar the world over while rural people embody the best in a national character and therefore more people should be encouraged to settle in rural areas (Aitkin 1985, Share 1995, Waterhouse 2002). Somewhat earlier small farmers were often 'represented as the backbone of the nation, credited with turning waste country into farmland' and even itinerant workers were valorised as 'tough and resilient, committed to mateship and defiant of authority' (Waterhouse 2002: 92-3).

The main constraint to supporting productive country residents and encouraging rural settlement was the concentration of wealth and power in the cities, exemplified in the primacy of Sydney by 1891 and into the twentieth century (Aitkin 1972, 1985). If Newcastle and Broken Hill were excluded, Sydney was at least 25 times larger than the next largest country town in 1891. By the turn of the century it was 50 times larger than Bathurst (the largest inland town at the time) and a decade later this ratio had increased to 70 times the largest inland rural settlement (Aitkin 1972). Not only rural people felt marginalised by Sydney, but rural townsfolk did too, and much support for the idea of countrymindedness came from people in country towns. In order to 'attract new settlers and to retain those already living in the country, conditions of life have to be made as attractive in country towns and on the farms as they are in the big cities' (Aitkin 1972: 18). This meant that from its inception the fledgling Country Party, representing rural and rural town interests against the expansion of, and appropriation by, Sydney interests, was 'fighting for "balanced development" against the forces of "big city domination"'(op cit: 19).

Although the rural population of Australia remained relatively constant at about two million people between 1901 and 1954, it declined from about 40 percent of the Australian population to around 21 percent during that time period (Davison 2005). Through rural–urban migration to coastal cities, the notion of countrymindedness as a national ideal may have been perpetuated through kinship between country and city residents (Aitkin 1985, Davison 2005), but the rise of coastal industrialisation, in cities like Newcastle and Wollongong, brought new sources of economic growth and symbols of modernity. Rapid population expansion after the Second World War, driven largely by immigration, was vital in reducing the significance of rural–urban links in political terms. Countrymindedness declined as a national ideology, but remained significant in parts of rural Australia, including both NSW and Queensland (Fitzgerald et al. 2009), notably at a personal level, as more recent generations of rural Australians have migrated from the country to the city for education or employment.

Countrymindedness has been criticised as an attempt to 'justify, enforce and naturalise relations of inequality and domination and to resist democracy' (Share 1995: 10), and as an ideology that sees rural areas, rural pursuits and rural people as different, and morally superior, to their city counterparts. Kapferer (1990: 104-5) argued that the separation of urban and rural interests 'provides an unquestioning and unquestioned foundation for the mythologising of Australianness as the embodiment of rural virtue, a virtue which is none other than an ideology produced in, by and for an urban industrial world'. Aitkin (1985), and more recently Lockie (2000), Smith (2004) and Davison (2005), have all argued that country people see themselves differently from city people, based on their life circumstances and the structure of economic organisation. While this ideological perspective is not as widespread as it was in the 1920s, it has not disappeared, despite departing from the centre stage of politics. Davison (2005: 52) suggested that 'by the 1970s, city folk were less inclined to concede the old moral and eugenic arguments in favour of country life. They no longer regarded the crop of country babies as biologically or morally superior to their own'. Three decades later city folk never even thought about such issues.

While city people no longer see rural people as virtuous and superior, many country people still see themselves as being different from city people, and city life as competitive, parasitical and inferior. A similar perspective on values and lifestyles was provided by one of the CW staff in 2008: 'I can't imagine not running into people you know down the street, and not being able to get to work in five minutes, and not being able to find a house at a reasonable price' – a vision of a country of abundant time and friendliness not overwhelmed by capitalist relationships. Urban residents would both challenge such assumptions as being invalid to urban life, and recognise that urban living comes at a financial cost but with distinct benefits. Nonetheless various people continue to subscribe to a loosely dichotomous view of the world, and such perspectives are a key theme in what CW proposes and many local councils support. While CW has undergone a name change, and experiences various challenges, the Expos are important events for many regional councils and state services – such as education and health – providing a chance to visit the city, meet each other, engage with urban residents and hopefully attract them to live in the country, especially their particular local area. Promoting country Australia takes multiple forms.

Whither the Country?

Particular assumptions and discourses are at the core of visions of rural and regional Australia, whether in the policies of the National Party (and of several independent members of parliament) that have diverged from those of its coalition partners, the primarily urban Liberal Party, or in the stories and advertisements of magazines such as *Live the Dream* and *Country Style*, and the national and local media of regional Australia (*The Land*, and *Stock and Land*). They also recur in the publications and pronouncements of organisations such as CW, that seek to promote rural and regional revival, and the councils that in turn support CW, and are the potential beneficiaries of its success. Even the Australian Labor Party, which despite a history in organised labour traditionally associated with urban and industrial concerns, has established a somewhat distinct Country Labor as the rural wing of the party.

Strategies for regional development that are overtly economic have ideological underpinnings, and successful regional strategies demand some obeisance to the perceived distinctiveness and needs of 'the bush', part of which is both retrospective and nostalgic. At the very least, as the opening quotation to this chapter indicates, regional Australia is 'different', and such differences may be both positive and negative. Even more pragmatically, parts of rural Australia are losing population and economic vibrancy, and practical policies are required for such regions, to boost economic development, and thus stimulate population growth or, conceivably, to 'smooth the dying pillow' of decline, a possibility occasionally hinted at (e.g.

Forth and Howell 2002). But ignoring a vast area is not an option. Pragmatic perspectives have been somewhat boosted by many of those leaving regional areas moving to larger cities, contributing in a small way to urban congestion and the concentration of economic activity there. While it is implausible that population decentralisation will significantly change metropolitan circumstances, it is one small factor encouraging regional growth, and the geographical re-balancing of development.

Over little more than two centuries of white settlement the fortunes of country towns have fluctuated, linked to global economic events, evolving resource appraisals, changes in technology and changed attitudes to rural and urban areas. While understandings of 'the country' may be simplified into a few dominant images, not only is there great diversity but ideas of 'the country' do not exist apart from the city. Populations have declined and aged in many rural and remote country areas but coastal regions have thrived. The next chapter develops this theme to explore various accounts of migration from cities, and how part of this movement has been an inland 'tree change', that may offer prospects for regional revival, rather than a coastal 'sea change', but both of which are inseparable from evolving perceptions and discourses of regional Australia.

Chapter 2
Leaving the City

While Sydney struggles to house its burgeoning numbers, western NSW is littered with ghostly abandoned farmhouses as, for two decades, the less profitable have sold out to neighbours or to the agri-corporations. The population has hollowed out, leaving fewer people to fight fires, help in floods, play cricket and keep the community humming (Jopson 2009: 9)

Urbanisation has never been either inevitable or without countervailing tendencies. Yet in a century where more than half the world's population is now urban, and that proportion increases annually, it may sometimes seem that way. But in every place and in every century there have been alternatives. Even before Victorian times and the industrialisation of urban England there were pressures for certain forms of population decentralisation, to much smaller and more elegantly planned townships such as David Dale's and Robert Owen's village of New Lanark near Glasgow, or the mid-nineteenth-century village of Saltaire north of Bradford. As the Victorian era came to a close the Arcadian visions of William Morris gave way to the Garden City vision of Ebenezer Howard who sought to combine the advantages of town and country. By contrast, and somewhat remarkably, Australia had no parallels and model communities were invariably urban (Freestone 1989) though, in the interwar years, soldier settlement resulted in returned soldiers being granted rural land, both as a reward for service and a means of settling seemingly underpopulated rural Australia.

In post-war Britain decentralisation policy was introduced as the Abercrombie Plan for Greater London in 1944, which promoted the expansion of some small towns and the creation of new towns, initially beyond a London greenbelt, and subsequently in northern England, Scotland and Wales (Thomas 1970). These plans built on the earlier ideas of Howard, which included favourable regard for working the soil and the benefits of country living, more than a flicker of countrymindedness. In Australia, the first push for planned decentralisation occurred in the 1970s under the Whitlam Labor Government and the Department of Urban and Regional Development (DURD), but despite earlier attempts at a green belt in Sydney, there were few parallels with the United Kingdom. Decentralisation occurred much later than in England because of the long era of Liberal Party political domination, and its unwillingness to become involved in planning.

The legacy of the belated decentralisation push is mixed. In 1972 it was proposed that Albury-Wodonga on the NSW-Victoria border was to have a population of 300,000 people by 2000, but in 1978 the target was revised to 150,000 (McManus 2005). While Albury-Wodonga is a thriving inland centre of 73,500 people and one of the three largest inland centres in Australia, it has done little to siphon population growth away from Australia's major cities. Other designated growth centres such as Bathurst-Orange (west of Sydney) and Monarto (south-east of Adelaide in SA) were even less successful, though each had good transport access and was not far from large cities.

The focus on encouraging population growth in inland Australia has been largely unsuccessful, and existing decentralisation is primarily due to market forces, particularly in Sydney where house prices and rentals discourage urban residence. Yet, as migration to Australia gathered pace in the 1980s and 1990s, metropolitan and coastal concentration merely intensified with migrants following their predecessors, in the largest coastal cities. A nation supposedly built on the sheep's back, and with a twentieth century national capital deliberately located inland, seemed to have resolutely turned its back on the rural and agricultural past for a coastal, urban and metropolitan future. It had become internationally renowned as a country where, even by the mid-1970s, over 85 percent of the population lived within 80 kilometres of the coast, with the rest thinly scattered over 85 percent of the continent: a nation living on the edge – on the balcony looking outwards (Drew 1994), and most of those on the edge were big city dwellers.

Second Homes and Coastal Growth

In most countries the 1950s and 1960s had marked a partial population turnaround as greater longevity and more affluent retirement took many retirees to small coastal towns beyond the large cities (Karn 1977). That same greater affluence, alongside greater mobility through the spread of car ownership, enabled the earliest phases of second home ownership to get underway which, while only temporarily and transiently shifting populations (through 'seasonal suburbanisation'), brought new economic life to many, mostly coastal centres (Clout 1974). Car ownership, affluence and improved highways stimulated retirement migration and the rapid growth of second, holiday homes in Australia, especially on the east coast. By the end of the 1960s their profusion characterised many NSW and Queensland landscapes, to the extent that 'the affluence-induced proliferation of homescapes has added a new dimension to the term "coastal erosion"' (Marsden 1969: 73).

While the rise of small coastal towns quickly came to typify Australia, long-distance commuting was rare other than in such exceptional cases as the Blue Mountains west of Sydney. However, it eventually became both a matter of choice and necessity, as metropolitan house prices encouraged and forced distant commuting. In many western countries similar affluence and improved transport brought the expansion of commuting significantly beyond suburbia to villages and towns removed from the metropolitan centres. However the geographical extent of such change was highly restricted (Pahl 1965, Connell 1978, Newby 1979) in a process of rural gentrification that often preceded its urban parallels. But, as in the title of Pahl's early monograph, *Urbs in Rure*, it had merely symbolic rural overtones. Whilst in England such commuting from many cities eventually reached a substantial part of the countryside, in Australia its extent was limited and most of the country was too remote, hence rural depopulation was largely unchecked by either increased numbers of second homes or commuting.

The 1973 Age of Aquarius festival brought a short phase of new age settlement and a limited rural revival in some areas. The small town of Nimbin where the festival was held, in a beautiful part of northern NSW, but where the old mainstays of the rural economy – bananas and dairying – were in steady decline, experienced new growth as alternative settlers took advantage of cheap urban and rural property prices and moved in, some to establish communes, others to live off the dole and some to commute to work nearby. A few other parts of Australia, including the south-west of WA, experienced similar changes but few were sustained over time, as communes failed and second generations preferred urban life.

The principal revival of regional areas was almost exclusively in coastal districts from second homes and retirement. New aspirations for increased leisure time coincided with greater affluence, marked by the transition from manufacturing to services as urban economies shifted focus. By the 1980s living on the edge had taken on new rural connotations, with second homes established in only a few relatively accessible inland areas such as Kangaroo Valley and the lower Hunter Valley, while retirement towns were almost exclusively coastal. Retirement usually took Australians northwards, but while many southerners moved to south-east Queensland distant migration was relatively unusual in Australia. This initially tentative process of counter-urbanisation had little effect on metropolitan areas as second homes were simply short-term residences, but as retirement migration became more important, and new purchasers spent longer periods of time in what gradually became first homes, so counter-urbanisation brought population shifts and real growth in coastal regions.

From its origins, the acquisition of second homes was touted as a lifestyle change – a chance to wind down and relax, to experience a smaller community, fresh air, sea breezes and open spaces and escape the pressures of urban life. In Europe this eventually led to a new genre of literature centred on bucolic village life, typically involving the purchase of rundown buildings, renovation, unusual encounters with the local population (not all of whom were amenable to foreign residents) and consumption, cooking and production of local food and wine. Foremost amongst this literature was Peter Mayle's *A Year in Provence* (1990), that touted the idiosyncratic virtues of southern France. A raft of other books on rural France soon followed, while in Italy the genre classic, and a relatively late arrival on the scene, was Frances Mayes' *Under the Tuscan Sun* (1996). The paperback's cover claims that she 'opens the door to a wondrous new world when she buys and restores an abandoned villa in the spectacular Tuscan countryside. In sensuous and evocative language she brings the reader along as she discovers the beauty and simplicity of life in Italy'. The inside cover queries 'Can we bear yet another book about buying and remodelling a tumble-down house in some sunny foreign country? The answer is a simple and unqualified yes'. Indeed both Mayle and Mayes produced sequels. While many second homes were utterly prosaic and indeed often wholly mass produced and modern, without need of renovation, and in places where fish and chips and beer were as common as slow local peasant food and wine, the visions, experiences and prose of Mayle, Mayes and a host of others nonetheless proved models for many who sought to follow. More than a hint of this experience trickled across the world to Australia.

By the end of the twentieth century a combination of increased affluence, earlier retirement, cheaper air transport and in some cases, quite ironically, 'white flight' (with metropolitan residents seeking to 'escape the flood' of new migrants into the cities) had resulted in both the acquisition of second homes and retirement migration extending much further afield. Areas such as Cape Verde and the Turkish Republic of North Cyprus had become new destinations for British migrants, and Bulgaria was marketed to skiers. New magazines appeared on British newsstands solely to market overseas property and lifestyle change. Second homes, property investment, relocation and retirement were bigger business than ever before, and rather paler echoes of each of these trends had begun to transform Australia.

Counter-urbanisation in Australia

By the 1970s the shift away from cities was well underway in Australia, though numbers were fewer, and second homes were more significant than relocation. Households were moving northwards from Victoria and, during the 1990s, Tasmania experienced actual population decline. The southern coast of NSW attracted Victorians, while the Brisbane–Gold Coast corridor became the fastest growing part of Australia, with Noosa and the Sunshine Coast north of Brisbane almost matching its speed of growth. More local moves out of Perth and Adelaide occurred in WA and SA. Counter-urbanisation was most evident from and around the largest cities, particularly Sydney, in what were called 'perimetropolitan zones' (Hugo 1994).

The consequence of movement out of the largest cities – Melbourne, Sydney and Brisbane – was that the east coast of Australia became transformed by population growth and redistribution, with only the most inaccessible areas exempt. The growth of tourism enhanced that transformation. Once sleepy fishing ports and occasional holiday resorts began to grow quickly. Symptomatic of such change was Byron Bay in northern NSW, an old whaling town where livelihoods that were centred on fishing and an abattoir in the 1970s gave way to the provision of tourism services and real estate. Tourism developed in a particularly distinctive way, centred on surfing, and 'alternative' cultures linked to the semi-tropical hinterland, westwards to Nimbin, and to a series of music festivals such as the East Coast Blues Festival, all of which brought economic growth, environmental change, the ascendance of a Green Party dominated council and disputes over the future direction and expansion of the now rapidly growing town (Gibson and Connell 2003). While Byron Bay was more dramatically transformed than most, rapid growth also utterly changed Airlie Beach, Noosa, Coffs Harbour, Kiama and a host of smaller towns.

The fastest twentieth century population increases in NSW and Queensland have all been in coastal areas, with Tweed, on the NSW north coast, growing at an annual rate of 2.6 percent, and large increases in Lake Macquarie, Port Macquarie-Hastings, Coffs Harbour, Byron Bay and Ballina. In Queensland growth has been even faster. Outside Brisbane the Sunshine Coast (2.9 percent), Gold Coast (2.7 percent) and Moreton Bay (3.4 percent) have grown particularly quickly while the northern coastal city of Cairns had an annual growth rate of 3.9 percent. Such places have sometimes been singled out as places overwhelmed by the speed of change. Environmental degradation of the rapidly changing Capricorn Coast around Yeppoon (Queensland) included hillside clearing (and erosion), high-rise buildings, alongside the 'commodification of elevated sea views', both of which processes 'corroded long-term residents' connection to place', with loss of green space and natural habitats, pollution of waterways, more costly water supply in difficult environmental conditions and the inability of small conservation groups to effectively challenge development trajectories (Danaher 2008). To many the Gold Coast of south-eastern Queensland became the dystopian example of over-development to be avoided at all costs.

This population turnaround was largely driven and facilitated by a combination of greater affluence and longevity, mobility, housing affordability and environmental amenity in regional areas and their converse in growing congestion pressures and declining amenity in urban areas. Rising affluence and improved regional infrastructure made relocation less challenging, and increased the distances over which it was feasible. The exact combination of factors varied from place to place and between different demographic and social groups. Indeed there have been attempts to distinguish between 'ex-urbanisation', where former urban residents move out to places of more pleasant environment but within regular access of the city, perhaps even for commuting (Burnley 1988, Burnley and Murphy 1995), 'displaced urbanisation' or 'welfare migration', associated with lower income households moving to rural areas in search of employment and cheap housing partly because of the transferability of welfare payments (Hugo and Bell 1998) – evident in the rise of relatively permanent coastal caravan parks (and 'deadheads' in Nimbin) – and 'anti-urbanisation', with the more complete rejection of an urban lifestyle and environment, including even a shift towards greater self-sufficiency (Sant and Simons 1993) or what later became 'downshifting'. Such distinctions tended to focus on changes and differences rather than either the continuity of lifestyles or a slow transition away from a perhaps more frenetic urban life. Ultimately counter-urbanisation was a rather nebulous phenomenon where there was some 'deconcentration' of population (and sometimes small industry) from large cities to rural areas, for a range of reasons, that usually involved some sense of a lifestyle change (Essex and Brown 1997), vague though this might be.

Outside Australia clarifications of counter-urbanisation are no better, and universally it is a vaguely defined term that serves as an umbrella for many local variations, though it usually implies a relatively wealthy middle-class group withdrawing from urban settings to pursue a somewhat different lifestyle in a rural location. Related terms include regeneration, dispersal, core-periphery migration and rural restructuring (Stockdale 2009), whether in Britain (Milbourne 1997, Lewis 1998, Paquette and Domon 2004), the United States (Ghose 2004) or Australia (Burnley and Murphy 2004, Costello 2007). More recently Hugo concluded that the social context of urban–rural migration patterns was scarcely novel since it was 'not so much a counter-urbanisation trend that is occurring but more a new, diffuse form of urbanisation' (2005: 78), an altogether more bland evolution towards the expansion of urban life and culture, and a parallel with Pahl's *Urbs in Rure*. Similarly Burnley and Murphy (2004) provided no real evidence of 'lifestyle' changes, or even psychological and cultural factors that had resulted in a desire for a changed lifestyle, and therefore argued that it made little sense to perceive counter-urbanisation as a radical break from past urban lives because 'much of the settlement in Australia associated with counter-urbanisation is residential-urban ... So the migrations are not counterurban: they are mostly associated with a non-metropolitan form of urbanisation' (Burnley and Murphy 2002: 151). O'Connor (2001) also argued both that it was the outcome of a desire for residential and commercial opportunities near metropolitan areas, and that it was not a cultural phenomenon, but more closely tied to employment issues.

Despite assertions that counter-urbanisation was merely a spatially fragmented variant on urbanisation, to a considerable extent it became conflated with counter-urbanism, a movement that deliberately rejected certain urban values – or merely urban pressures – in favour of a more relaxed and small-scale life style: simply conceptualised in the mid-1970s in the United States as a yearning for the countryside (Berry 1976). In England, where similar sentiments were expressed, lifestyle could be seen as 'seeking a new direction, a change from their daily routine, career and commuting, and the prospect of a less pressurised life ... [and] escaping from the rat race' juxtaposed against urban 'crime, vandalism and insufficient open space to bring up a family' (Bolton and Chalkley 1990: 38, cf. Jetzkowitz et al. 2007). Such broadly generic characteristics were widespread, but similar expressions of anti-urbanism, in any ideological or even practical sense, were limited, despite much literature and many migrants touting the distinct virtues of life in smaller more rural places.

In Australia counter-urbanisation was similarly regarded as a rejection of oppressive urban life, enabled by lower housing costs (Sant and Simons 1993, Walmsley et al. 1998, Burnley and Murphy 2004), yet few studies produced much evidence of anti-urbanism or of the demand for an alternative. Typically the much cited lifestyle reasons were vague, such as the 'relaxing lifestyle', 'pleasant climate' and 'attractiveness of the coastal location' that were the three key factors that drew newcomers to Byron Bay and Ballina on the NSW north coast, along with a 'good place to raise a family' (Essex and Brown 1997, Gurran and Blakely 2007). Likewise, beyond the fringes of Adelaide, 'counterurbanites move ... in pursuit of quiet, peri-urban lifestyle' (Fisher 2003: 563). So-called 'downshifting', with deliberate attempts to reduce consumption, attracted some interest but was no explicit element of counter-urbanisation, unlike in Scandinavia (Hamilton and Mail 2003, Lindgren 2003). Salt however perceived more radical change: a 'big shift' in both geography and aspirations in the movement of people 'wanting a simpler life in a pleasant town with all the amenities not too far from their interests in the city', which was the result of a fundamental shift in Australian culture, away from urban culture to a 'culture of the beach' (2001: 23). He retained the same view years later, over parallel movements inland:

> Tree changes still fulfil the same need for a lifestyle – you just don't get the blue view. The principle of a sea or tree change is still the same: to slow down your life and enjoy the environmental amenity ... a difference in mentality is part of the attraction for city people coming to Sea or Tree Change locations ... coming out of the anonymity of the big city, and into a very embracing and intimate local community – with real quality of life (quoted in *Live the Dream*, 2008: 10, 12).

What this meant in practice, and what constituted 'good places', new 'lifestyles' and better 'quality of life' were rarely taken further. Many migrants may have been just as unsure. There was little to indicate what such distinctions as 'urban culture' and 'beach culture' entailed, or why they could not be combined, let alone whether there had been any real shift in orientation.

Despite much of the literature on counter-urbanisation being focused on changing lifestyles, it was rarely explained exactly what this meant, what lifestyles had actually changed, whether this was deliberate (or even enforced by more limited facilities), or whether change was simply some degree of 'slowing down' as households aged and became couples again. How much scope for or interest in change did most people have? It was often simply implicit that aging must drive changes, and perhaps new values, and that moving to new environments must result in change. On the other hand, for all those not driven purely by economic phenomena it was surely axiomatic that some change was at least intended, especially in the shift towards consumption-based societies where 'lifestyles' became cultural markers. Arguments from new residents that they had slowed down and changed their lives in positive ways, as might generally have been expected (certainly in survey responses), matched the invocations of real estate agencies and other boosters of regional places that changes were possible and invaluable. Counter-urbanisation at least provided a rationale and context for change.

While much counter-urbanisation was retirement migration, with couples seeking a more relaxing place and pace, a substantial proportion was also of families seeking to make some sort of lifestyle change decades before retirement, and even of the unemployed seeking to hang out in more idyllic locations, where the weather was warmer and surf abounded. While many small coastal towns nonetheless became characterised somewhat pejoratively as mere retirement towns (or, worse, 'god's waiting rooms'), in many cases, as at Merimbula on the south coast of NSW, households were moving there long before retirement, sometimes after a period of regular holidaying or after having a second home there, because employment was available (Cairns 1991). Elsewhere in the immediate environs of the largest cities, overspill was forced on people by market forces, because they were unable to afford (sub)urban residence (Gurran 2008). At the very least there has long been a considerable diversity in counter-urbanisation, all hinting at issues and means for stimulating relocation to particular places.

Setting the Mood for Sea Change

By the end of the twentieth century this population turnaround, with its emphasis on coastal areas, was beginning to be referred to as sea change, again with implications that relocation was associated with lifestyle changes. Its country cousin, tree change, was still invisible, as inland Australia remained largely apart from counter-urbanisation and populations continued to stagnate or decline. The idea of sea change has long passed into popular usage in Australia, being widely taken up by publicists and academics (Salt 2001, Burnley and Murphy 2004), and constantly regurgitated in the media, especially after the Australian Broadcasting Commission (ABC)'s late-1990s Sunday night drama series *Sea Change*, which ran through three series from 1998 to 2000. Set in the fictitious Victorian coastal township of Pearl Bay, loosely modelled on and partly filmed in Barwon Heads, 100 kilometres south-west of Melbourne, it starred a female lawyer who had moved to the town, after her marriage broke up, in search of a quieter life style. Pervasive themes throughout the series involved the tensions and differences between 'outsiders' and 'townsfolk' and the challenges of settling in and, for some of the characters, returning to a small town. The dominant demographic structure of viewers was educated professional households between 40 and 65. Its theme song set the mood:

> Don't wanna live in the city, my friends tell me I'm changin',
> The smell of salty air, is what I'm chasin'.
> You probably think I'm mad, but it feels good to me,
> Cos from now on I'll live as close as I can to the sea
>
> I don't know why I'm going through the seachange
> I'm reaching out to the sky for a seachange
>
> If life begins at 40, well I can't be bothered waitin'
> My therapy is goin' nowhere, there's nothing I've been takin'
> You probably think I'm mad, but it feels good to me,
> Cos from now on I'll live as close as I can to the sea
>
> Sky is blue (the sky is blue) and I'm going to (and I'm going to) a seachange (a seachange)
> The time is right, now I'm going through a seachange[1]

---

1   'Seachange'. Words and music by Richard Pleasance. © Universal Music Publishing Pty Ltd. All rights reserved. International copyright secured. Reprinted with permission.

The original *Sea Change* had a real impact on migration, above all in enthusing a new group of people about the possibilities for change. At the height of its popularity, its producer Deb Gibson noted on the website that 'So many of us are longing to escape from the urban madness. We want to be able to relate to our families, to know our neighbours by their first names, and find values that are more enduring than a microwave dinner'. It directly led to a substantial tourist boom in Barwon Heads, which later gained new residents and increased real estate prices (Beeton 2001, 2003). Its population increased from 2,128 in 1996 to 2994 a decade later.

The considerable success of *Sea Change* brought other series along similar lines. *Bed of Roses* (which also starred Kerry Armstrong, a prominent cast member from *Sea Change*) examined 'enforced sea change' as the lead character lost her husband and income simultaneously and was forced to move away from the city. Two series of *East of Everything*, set in the fictitious coastal town of Broken Bay (a combination of Broken Head and Byron Bay, and filmed in the latter town), and with the same producer as *Sea Change*, continued much the same themes and evocative scenery. The 'reality' show, *The Farmer Wants a Wife*, challenged urban women to experience life on Australian farms and perhaps ultimately become a farmer's wife. *The Real Seachange*, a second reality television series, with two separate seasons in 2006 and 2008, followed families, couples and single people who left large cities in searching for a better life in various locations including inland centres. The series was narrated by the actor John Howard, one of the stars of the fictional series and previously a star in *Always Greener*, a program about urban and rural role reversals.

Collectively they presented a picture of a rather different, more quirky, perhaps more engaging, but also intermittently challenging, lifestyle away from large cities. At the very least they constantly raised the idea of a movement towards a different kind of regional lifestyle. They also coincided with a penchant for numerous television series, featuring most Jane Austen novels and others of the same genre and period, portraying glimpses of a quieter past of genteel manners, elegant houses and ordered country life, where people knew their place in an all embracing community. In very different ways television screens purveyed retrospective and nostalgic visions of alternative lifestyles, in rural settings.

Tuscany under Different Stars?

While the books of Frances Mayes and Peter Mayle, and a host of sequels, parallels and films and television programmes, boosted European fascination with moving to different regions and developing new lifestyles in locations such as Tuscany, in Australia where modern settlement was scarcely centuries old there were no comparable peasant regions or tumbledown houses to renovate in quaint villages. Australia, a land of small towns and scattered farmhouses, certainly had no Tuscan or Provencal villages.

Yet Mayes and Mayle were as popular in Australia as they were in the northern hemisphere and there were echoes of the changes that they ignited in Europe. Farmhouses could be renovated, but only those in more accessible regions, such as in the river valleys north of Sydney, attracted much interest. Many of Australia's wine regions in the 1990s however, began to take on more evidently European images, which even in areas like the Barossa Valley of SA, with a strong European heritage, had tended to lie dormant; Tuscan farmhouses sprung up in the Hunter Valley and the marketing of wine, cheese and olives took on cosmopolitan overtones, and contributed to the reconceptualisation of rural landscapes. Elsewhere, as in parts of rural WA, farming initiatives included herbs, blueberries, Chinese chestnuts, mead, llamas, rabbits, deer, emus and ostriches (Curry et al. 2001) as extensive conventional wheat and grazing lands gave way to more intensive and exotic systems, that smacked of taste and distinction.

The emergence of the magazine *Australian Gourmet Traveller* in 1989 marked a new stage in the conceptualisation of countryside. While many features were devoted to the pleasures of international travel and food and wine consumption in Europe, Asia and elsewhere, a regular focus on regional Australia enabled it to be perceived in a similar way. Good food and wine and enjoyable travelling experiences might also be found in Australia. The perception of regional Australia in a new, somewhat elitist, light, and not as a backwater to urban dynamism, was further and more comprehensively enhanced with the arrival of a similarly glossy magazine *Australian Country Style* in 1998 (from 2008 simply *Country Style*) that took on much of the aura of similar English magazines and specifically focused on 'the Country' as a place of elegant houses, farmhouses and gardens, good food and wine, fashion and enjoyable travelling; above all it was inherently a place in which to live a good life. As one letter to the Editor observed

> Earlier this year we moved from the concrete jungle of Sydney to a piece of
> paradise in north east Tasmania. Many months later I happily sit in my newly
> decorated parlour, gazing at the to-die-for view of green pastures and distant
> mountains, sipping tea and leafing through my *Country Style* magazines that I
> have collected from the very beginning. Ah yes ... this is how life is meant to
> be! (*Country Style*, August 2009)

On the back of perceived interest in rural living, and with the support of CW, a new magazine, *Live the Dream: Sea and Tree Change Australia* emerged at the end of 2007, though it took more than a year before moving on to its second issue. Perpetuating the rural–urban dichotomy, a theme that pervaded much of the popular literature, the editor's letter in the first edition of this magazine began: 'The city and the bush: it seems there have always been two Australias' (Turner 2008). *Live the Dream* was deliberately designed to extol the virtues of regional residence and was a more middle class version of CW's own broad strategy (Chapter 3).

Multiple magazine features have taken on similar themes as lifestyle magazines, like *Country Style*, have become more widespread, matching rising affluence.

South Coast Lifestyle, for the south coast of NSW, and the national *Coast and Country* pursued the same themes with the same glossy format. Alongside other newspaper stories, especially in the travel supplements of weekend editions, and features in various magazines and colour supplements, regional Australia has taken on a new guise, no longer ignored and where elite European themes have become more pervasive. While feature stories may be relatively subtle, advertisements in such magazine do not seek to be, as in a tourist advertisement for the Limestone Coast of SA

> I might not be back on Monday. Time moves a little slower ... when grazing the menu. The days meld into one another, the wine seems endless and the food actually tastes better. Indulge the senses in the Limestone Coast where world class wine blends brilliantly with fresh local produce (*Country Style*, August 2009).

The Limestone Coast, like 'Hunter Valley Wine Country', is a relatively recent construction, positioning rural Australia in a particularly attractive way, while brochures encouraging domestic tourism naturally took similar perspectives. Regional Australia was to be consumed, to be enjoyed and to be populated.

In Australia, the former television personality Patrice Newell's account of relocating to and revitalising a rural property in the Hunter valley northwest of Sydney (aided and abetted by her partner, and fellow media personality, Philip Adams), was influential. One website (Biome) advertised the book *The Olive Grove* (2000) as

> Patrice Newell's tree-change book. The story of leaving television, becoming a biodynamic farmer and the struggle to plant the first olive grove. When Patrice Newell left a highly charged urban life to live in the country, friends warned her she'd be bored in no time. Now she is doing more in a day than most of us do in a week ...

She called it her 'sea change book', as she simultaneously left urban life and a modelling and broadcasting career, and it too had sequels including *The River* (2003) and *Ten Thousand Acres. A Love Story* (2006). Similar books included Annette Hughes' *Art Life Chooks. Learning to leave the city and love the country* (2008) which similarly chronicled a move from central Sydney to a farm in Queensland, and reflected on the significance of seasons and the various travails of agricultural life. Each was set in areas of considerable beauty, within easy reach of urban life and distant from featureless plains. More important than any single component of these changed media perceptions and literary evocations was the evolution of a perception that, if scarcely Provence or Tuscany, regional Australia had its charms and was worth visiting and even living in. The television drama *Sea Change*, and its sequels, were in fact simply rather proletarian versions of *Under the Tuscan Sun*, but, as much as anything, they played a major part in re-placing regional Australia on a metropolitan map, though that map retained a largely coastal bias. In a country whose unofficial national anthem sung of billabongs, sheep and swagmen, where 'drover's dog' was an epithet vested on inadequate politicians, 'a wide brown land' was part of the most famous national poem, and whose most famous films and paintings invariably depicted the inland, the inland had otherwise become curiously absent from everyday life. However the process of reconceptualisation had begun.

Tree Change

While elite publications belatedly focused on inland Australia, Australia's population remained thoroughly coastal, whether in expanding cities or coastal growth areas. Sea change might have become fashionable by the start of the present century but tree change – any comparable decentralisation inland – remained conspicuous by its absence, and the term only gradually entered the lexicon in the early 2000s. While some inland areas had grown, whether new agricultural areas centred on larger towns like Mudgee and Orange, or in the booming mining areas of WA and Queensland, they had not experienced the kinds of population movements that had so transformed coastal Australia. Even the steady shift to a post-productivist countryside, as traditional agricultural resource production declined, and property prices stagnated or fell while those of urban Australia were soaring, brought minimal change.

By the start of the present century there was at least some sense of movement inland, but again mostly associated with a number of favoured locations. Salt was using the term 'tree change' at least as early as 2004 and arguing that many tree changers were looking for 'arcadia meets urban chic' and suggested that they were finding it in such towns as Oberon, Bathurst and Orange (quoted in Pryor and Lewis 2004: 25). While he had detected a 'cultural shift' that had brought on the sea change, the tree change seemed to offer no parallels and he simply conceded that 'not everyone likes the beach' (Salt 2006: 3). By contrast Burnley and Murphy (2004) argued that growth in regional areas away from the coast, such as Dubbo and Tamworth, resulted primarily from their being regional service centres rather than through amenity led migration. Qualitatively and quantitatively, as growth was slower, inland Australia was different from the coast. Burnley and Murphy did however note that movement inland might increase

> If a person's notion of amenity is broader than a beach and a suntan, then non-coastal areas have a potential that a substantial and perhaps increasing *minority* of sea changers recognises ... This potential will perhaps be enhanced by the astronomical housing prices in many coastal localities ... Various processes may well see accelerating diversion of growth away from the coast. Apart from rising land prices ... other factors that may steer growth to the inland include a reaction against the scale of essentially suburban-style development on the

coast; expansion in a nostalgic re-identification with inland/bush myth by baby
boomers looking for ex-metropolitan places to live and the decline of small
towns, resulting in a stock of low cost housing (2004: 47-8; our italics)

At least three things were clear. Tree change was yet to become significant, and was
more likely to be driven by coastal failings (indeed a small scale version of what
had resulted in sea changers leaving metropolitan centres for lifestyle reasons)
rather than any innate attractions of the inland; coastal migration still had the edge
over moving inland and this was unlikely to change quickly; and amenity might
be of no great consequence. Despite Australian folklore about living in the bush,
how many baby boomers might become nostalgic about the bush was unclear.
Neither Burnley and Murphy nor Salt had surveyed inland change hence the extent
to which any 'tree change' was a function of cultural shifts or an economic focus
on employment and housing was untested. Over time there was a slowly emerging
perception that sea change has been so rapid that coastal towns were becoming
crowded and losing their sense of character and community as population changed
rapidly, so that 'retirees were no longer looking for a sea change but preferred a
"treechange" back to the country where people were still friendly' (Duggan 2005:
16). Tree change might not therefore have been necessarily preferred.

    Salt has continued to regard tree change (albeit without supporting data) as
both a cultural phenomenon and also the poor person's sea change:

> Yes there's tree-change but, let's face it, tree change is for sea-changers who
> can't afford pounding surf views. There's always that niggling suspicion that
> deep down tree-changers hanker for a harbour view. A 'green scene' is not the
> same as a 'blue view'. In either case the best tree-change towns are sited by
> a river or lake. Victoria's Daylesford – the tree change movement's exemplar
> – is defined by its lake and by the Lakehouse restaurant. There will never be a
> fashionable tree-change town with a 'hatted' restaurant in any of the flat desolate
> villages of the Australian wheatbelt: wrong karma (Salt 2009a: 2).

But this is purely an elitist perception, modelled on *Country Style*, yet somehow
combined with greater value for money inland. Salt has gone further, arguing that
we need to 'tighten the definition' of tree change, by refining his description:

> Ideally a tree change town is within striking distance [250 kms] of a capital city;
> is placed within rolling green hills (lightly wooded with eucalypts is preferable);
> has a heritage-listed main street with Australiana architecture; has recently
> attracted city sophisticates who are busy establishing a local gourmet food and
> wine industry; and, for the piece de resistance, has a genuine link to real celebrity
> … Selected towns must have a modest median income, low unemployment and
> not have too many old people or young kids [or] an insufficient mix of overseas
> – as opposed to Australia-born (2009b: 4).

Elitism is retained in a vision of Australian townships that excludes not merely the relatively remote but a host of places. Daylesford is unique, both as a spa town and an emerging gay centre, while larger towns like Orange and Mudgee in NSW are similar to WA's small town of Denmark, marked by 'elite consumption, up-market resorts [and hotels], specialised antique dealers, gourmet foods and wines and rural investments such as wineries' (Curry et al. 2001: 122). While they may fit Salt's perception, this version of tree change is atypical of much of rural change and those few places where this has occurred have no great interest in boosting migration further.

Despite Salt's lack of interest in attaching a definitive cultural designation to tree change, and Burnley and Murphy's argument that inland population change was primarily an economic phenomenon, over time 'tree change' became commonly used simply to refer to an inland version of sea change – perhaps a poor person's alternative, since property was much cheaper inland, or conceivably a desire for a lifestyle that was more easily possible inland (which usually meant the wish to live 'on acreage' – a small rural property of some kind). Such desires have a legacy in the 'hobby farms' of 'Pitt Street farmers' drawn to the country from the eponymous main street of Sydney, to avoid taxation, but whose ambitions and lifestyles were not those of the bulk of rural residents. However, without definitive studies, there were few indications of where tree change had occurred – and by whom – so that tree change tended to be defined and described as merely a geographical counterpoint to sea change.

By contrast, in Britain, where inland migration was much more substantial, its cultural significance had even been argued to be so strong that it was described as part of the creation of a neo-tribal identity, linked to the acquisition of a 'rural lifestyle' by new rural residents, as a radical response to contemporary urban society and the experience of capitalism (Halfacree 1997). Even in Britain such a notion was far-fetched, with few new rural residents detached from either urban life or capitalism, most anxious to preserve the gains that their rural investment had endowed, and few in harmony or even communication with longer established rural residents (Pahl 1965, Connell 1972, 1978). It had even less resonance elsewhere, although in New Zealand, a country with considerable affinities to Australia, migrants to rural areas were drawn to visions of fertile fields and rural idyll: a place of harmonious social relationships, security, community life and a more suitable environment for children. Material considerations took second place (Swaffield and Fairweather 1998). Whether similar idyllic notions characterise Australia remains to be seen.

Australia has experienced a form of counter-urbanisation, primarily coastal, as unbalanced as any in the world. Despite images of national identity that attach significance to the bush relatively few Australians are familiar with or have any experience of life in regional Australia. Moreover images of 'the bush' in urban Australia are linked to flies, and rather more lethal animals, heat and bushfires, and to a series of films from the 1960s classic *Wake in Fright* onwards, which challenge any idyllic notions of rural life. Transforming perceptions remains a challenge, especially in a nation of immigrants – a quarter of Australians (4.4 million people at the 2006 Census) were born overseas – that have come to the coast from overseas rather than from the inland. Much analysis of counter-urbanisation has been centred on changes in lifestyle, however vague, and even on perceptions of countryside that continue to paint rural life in bucolic tones, reminiscent of an all too elusive English countryside of green fields, woodlands and country cottages. In various contexts migrants to rural areas have seen this as somehow seeking to 'connect', either with a place of their youth or with an ethos that such places, in another sense of the past, were how the world should be, after urban life had become difficult and even threatening. Whether that is equally true of Australia, or whether rather more prosaic reasons have resulted in tree change, remains to be seen. It is necessary therefore to examine why people have moved to inland Australia, what processes were involved, and what that might mean for them and the places to which they moved. Identifying what if any new lifestyles are offered in inland regional Australia, and how these might be mobilised to attract new residents, is the critical question that underpins the objectives of CW and regional councils. In short, how can regional Australia be marketed to ensure that tree change ensues?

Chapter 3

Country Week

Whichever government wins the forthcoming election, can – or should – it do anything to support small rural towns and roll back rural deprivation and decline? Should Australians value their rural areas – or are they simply an economic resource that has had its day? (Digby 2004: 7).

Despite both long-distance commuting and counter-urbanisation, Australia remains a developed nation where cities are still growing at the expense of the country, and there is an uneven rural population decline. Counter-urbanisation has been a spontaneous process with households and individuals making personal decisions seemingly uninfluenced by advertising (other than for residential estates) and largely to coastal regions that have no need to encourage migration or promote themselves. This resulted in a largely middle–class counter-urbanisation, with fewer working class or 'blue collar' migrants. With rare exceptions it brought no significant change to inland regions where population decline was as evident as growth. In recently promoting counter-urbanisation, Australia is one of very few countries promoting institutional attempts to 'sell' rural and regional areas to urban households, and encourage urban–rural migration.

While state and federal government bodies have sought to promote regional development, through various economic strategies (Beer 2006), Australia has been unusual in promoting the virtues of regional residence to ordinary households. Despite some state initiatives in Victoria and Queensland, the principle vehicle for the recent promotion of regional Australia has been Country Week (CW), a private sector organisation that has promoted regional NSW and Queensland through annual three day Expos, where councils, government departments and others publicise the advantages of regional residence outside the largest cities. CW is a rare example of collective place-marketing to promote inland Australian towns, whether to secure growth, reverse population decline (in some cases) or fill employment vacancies in booming resource towns. This chapter explores participation, particularly of local authorities, in CW, focusing on the reasons they attend, their expectations and strategies for success, and the organisation of the Expos. While some councils have other promotional activities, for many the annual Expo is the main event in which they participate. For some councils it is the only event.

Country Week

The idea of CW began in the northern tablelands of NSW, a stronghold of the National Party (the former Country Party) and a region dominated by upland grazing. New England was the political base of the former National Party leader, Ian Sinclair, its member of parliament for 35 years until the late 1990s. At the centre of New England is the university town of Armidale, and a major regional centre, Tamworth, but with several smaller towns such as Glen Innes, Walcha and Uralla that were struggling to survive economically at the end of the twentieth century. Only Tamworth has grown significantly since the 1990s.

CW was established in 2004 by an Armidale businessman, Peter Bailey, who had been the National Party candidate in the 2003 NSW state election for the seat of Northern Tablelands. Bailey became the Chief Executive Officer of CW, which began by organising an Expo to highlight the attractions of living in rural NSW. The first CW Expo in 2004 was financially supported by the NSW Government, which contributed $100,000. Subsequent sponsors varied and included both state governments, and various private organisations such as the hotel chain, Mercure, the New England Credit Union, Country Range Farming and Adam Purcell Meat. In 2009 CW changed its name to become the Foundation for Regional Development Ltd.

CW has long advertised itself as promoting the 'Sea and Tree Change Expo' but its focus has primarily been on inland areas. Its budget comes in part from Australian state governments, either directly or through the ongoing support of government departments attending as participants, and from the various council participants. Both in being a private sector organisation (albeit with significant public funding) and being oriented to ordinary households rather than to broad notions of regional development or the transfer of firms and government departments, CW is quite distinctive. Only apparently in Sweden have similar structures and strategies marketed places directly to households (Niedomysl 2004), but not at an Expo, in contrast to more diffuse forms of place branding that focus on passive advertising, oriented primarily at economic development and tourism (Mayes 2008). Central to CW's campaign was the concept of local councils and other state bodies paying to promote rural and regional areas in the two state capitals, to address what they perceived as misconceptions about the country and to promote the specific advantages of their own local areas.

The CW Expos have been state-bounded, promoting only within each state, despite the presence of Tenterfield (a northern NSW town) at the 2008 Queensland CW. Consequently Victoria and Tasmania, which would like to attract people southwards from NSW, and SA, which would like to attract people westwards, cannot participate in the Expos. Nor have individual towns normally crossed state borders to attend, because that risks state funding support. One outcome was that CW was unable to get publicity in popular national magazines such as *New Idea* and *Woman's Day* for a purely state based activity, which limited its ability to reach a broader audience.

Other Australian state governments, notably Victoria and Queensland, have attempted to promote regional parts of their state. The Victorian campaign *Make it Happen in Provincial Victoria* was launched in 2003 by the state government in partnership with the 48 rural and regional councils. The campaign had similar aims to CW, notably to correct 'misconceptions', encourage city dwellers to move there and unite rural and regional Victoria under the banner 'Provincial Victoria'. Its campaign featured 'real people sharing their positive experiences in regional Victoria. It presents regional Victoria as an attractive alternative to Melbourne and reinforces to young people that they don't need to move to the city to have a bright future' (Victorian Government 2007). Queensland's program, *Blueprint for the Bush*, was intended to be a ten year program involving the Queensland Government, the Local Government Association of Queensland and Agforce (an agricultural lobbying group) that was launched in June 2006. CW's activities in Queensland, specifically the Expo in Brisbane, dovetailed with this program because the aims were complementary – to change negative perceptions about rural and regional locations and to encourage migration, particularly of skilled labour, to these areas.

Australian regional areas have long been disappointed with what they perceived as, at best, weak support for regional development strategies from both national and state governments. Tourism strategies, infrastructure projects (especially roads and telecommunications) all seemed to be ignored. While rural people normally lament urban inattention, and perceptions of urban bias are highly likely when most political support for state governments comes from the state capitals where a large proportion of the state's population live and their issues are daily visible, support for CW represented support for an organisation that placed regional areas first. State governments have given funding support to CW, and the Expos are invariably opened by leading politicians, including national and state ministers. While they have primarily been from the Australian Labor Party, that has been in power in both Queensland and NSW for the first decade of the present century, they are supported by all parties who can see advantages (and political expediency) in boosting regional centres and reducing urban congestion.

CW has survived initial tests of endurance and become an important event for many regional councils – a chance to visit the city, meet each other, engage with city people and attract some to visit and live in the country. CW emerged out of a sense of countrymindedness, where greater priority to regional Australia was important for national development and the realisation of national ideals. Councils participated where they could see real benefits from increased populations of particular kinds, or spin-off benefits from tourism, that would contribute to economic growth, and where crude cost-benefit analysis suggested it was worthwhile. This often meant the absence of more impoverished and remote areas from CW. Participation enhanced a sense of local community in bringing together sometimes disparate groups, but all of which shared some notion of the need for local development. Over time the majority of councils recognised that the most important stimuli to attract new residents were employment and housing, while services were crucial, whether for older people or families with children. Whilst aesthetic considerations were important, and linked into widespread but diffuse notions of lifestyle, they were gradually recognised as secondary, despite the necessity for lifestyle to be a selling point. How exactly various councils have sought to emphasise the attractions of their regions, compete with a host of others, and meet the expectations of visitors to the Expos, is described in this and the next two chapters.

Country Week and Countrymindedness

CW sought to portray and offer a positive countryside different from the perceptions that some urban residents were presumed to have, seemingly centred on what Peter Bailey habitually described as 'drought and dead sheep in the dam'. It promoted regional Australia as a place (or places) with job opportunities and affordable housing, alongside qualities that were perceived to be absent in large cities such as Brisbane and Sydney: community, peacefulness, lack of congestion and pollution, security and so on. Such positive regional portrayals are evident not only in CW's own publications, and those of local councils, as could easily be predicted, but in local newspapers from the regions and various magazines, where a series of themes constantly recurred (Chapter 8).

Although CW turned to the largest cities to re-populate regional Australia, an inherent anti-urbanism underpinned marketing strategies. Four strands of country-mindedness were evident in CW literature and at the annual Expos: an anti-urban bias, a pro-country bias, a concern about 'imbalance' and a desire to promote national development. When asked if countrymindedness was a motivating force for CW, one CW official responded: 'Oh, very definitely. I grew up in the country ... most of the town's economic base was based on the rural surrounds ... those properties employed two or three or four families. I don't know anybody who employs a family anymore'. In some respects the essence of both the rationale behind CW and the basic idea of countrymindedness is summarised in its 2008 webpage seeking to attract visitors to the Expos:

> Every day it is demonstrated to us why it is becoming more and more difficult to live in our major cities. Maddening congestion, commute times becoming ridiculous, traffic that seems to increase on a daily basis, fears for our own and our families' safety and on top of all of this, the near impossibility of owning our own home in the foreseeable future, and the increasing difficulty of even renting accommodation. The cost squeeze facing capital city based families continues to tighten. In Sydney, families face the situation of earning more money but actually having less to spend. But if you have taken the time to read this, you already know it. What you might not be aware of is that there is an alternative. Never has the time been so right to make the move to a country or regional centre.

The webpage for the 2008 Queensland Expo was even more pointed:

> Country Week is your opportunity to find out about the virtues of living and working in country and regional Queensland but whatever you do, don't make

the mistake of thinking of country and regional areas as 'the bush'. Sure, that wonderful part of our state still exists but Queensland, outside of Brisbane, is much more than that. Our State boasts many vibrant and sophisticated regional cities where residents enjoy the facilities and opportunities available in any capital city without the enormous mortgage payments, huge fuel bills and overcrowding and commute times that seem to grow by the day. It also boasts beautiful, clean and uncluttered coastline and beaches and smaller country towns where a real sense of community and caring exists and where affordable and easy living is available. Maybe it's time to ask ourselves these questions: ' *How much better off am I now than I was three years ago?* 'and ' *What is going to change in three more years if I stay where I am?* '. These can sometimes be confronting questions but as our lifestyles disintegrate in Brisbane it is fast becoming decision time if we want to change the situation. It's time to make the move to a country or regional centre and take advantage of the wonderful opportunities and lifestyle that is available outside of Brisbane.

These themes of positive regional experiences and a problematic urban life constantly recurred, as CW sought to emphasise the virtues of regional life against a backdrop of agricultural decline, focusing on employment opportunities and housing prices, designed to attract a working population less able to afford the costs of big-city living. Such themes effectively de-linked CW from agriculture. As a CW staff member stated:

> It is something that we have very deliberately done, not because we have any issues with agriculture, but we want to sell to city people that there is a lifestyle out there that is not necessarily rural – it can be rural if you want it to be, but we can provide you with sophisticated towns with good coffee, medical facilities, education, whatever you want is available somewhere out in the country.

CW thus added a distinctive dimension to counter-urbanisation in not exclusively promoting lifestyle changes (or tourism), but stressing practical amenities that would attract people to rural areas by being the converse of issues that might be causing challenges in the city. It was not however promoting 'welfare migration' from urban to rural areas, simply because these areas were more affordable. CW and the councils wanted 'working' people, not people who were welfare-dependent.

However many of the publications of CW and the councils emphasised lifestyle considerations, as the previous quotation and the websites suggest. The success stories (and title) of *Live the Dream* (a title taken from a constant CW epithet), and in some local newspapers, were at some remove from more prosaic attempts to encourage the migration of 'battlers', with valued skills and with young families, where jobs and housing were necessary and lifestyle goals somewhat secondary (see Chapter 8). ('Battlers', a term beloved by the former Liberal Prime Minister, John Howard, were said to be semi-skilled 'working families' – an even more popular phrase – usually occupying more working class areas of cities, for whom a better deal was needed. In earlier times they tended to be working class rural folk, doing it tough, but surviving despite the vagaries of nature). The orientation of CW was both ambiguous and ambivalent, seeking all who might migrate, recognising both a need for tradespeople and a desire for the free-spending middle class, and thus offering a smorgasbord of possibilities, including parts of the NSW and the Queensland coast.

The recurrent dichotomy between positive regional experiences and problematic city life raises questions about rural identity, the understanding of socio-economic status in rural migration, and especially at what scale urban life may not be problematic but positive. Even the name 'Country Week' (that organisers long debated) suggests a certain bucolic and perhaps even British notion of countryside, where agriculture is dominant, and urban life absent. But CW dismissed the agricultural focus in favour of an encompassing vision, where urban life is evident. (After the 2008 Queensland Expo, CW changed its principal slogan and name from the CW Expo to the Country and Regional Living Expo, in response to the perception in Queensland that 'Country' gave too much emphasis to rurality and lifestyle and not enough to crucial issues of employment). What is it about rural living that can provide a counterpoint to the opportunities offered in the city? Which segments of the urban population are more likely to recognise the advantages of living in rural areas, be able to adjust to any disadvantages and perceive living in rural areas to be better overall than living in the city?

Despite urban–rural migration being slight, it has been widely argued that the demand for a move is substantial. On the basis of a *Sydney Morning Herald* opinion poll undertaken after the Queensland on Show Expo in Sydney (an alternative state government Expo, along similar lines to those of CW but touting inter-state mobility), it was said that 'one in five Sydneysiders are thinking of leaving because of high living costs, better job opportunities, or traffic congestion and overcrowding' (Creagh and Nixon 2008: 1). What is central to all studies of migration, however, is that intent is never reality, and while many do consider changing location at some times(s) in their lives, few actually move. Probably the greatest challenge for CW and the councils is to convince potential residents and workers to translate interest into intent and ultimately into mobility.

The Country Week Expo Experience

The central element of CW is the annual Expo which (ironically) lasts three days, whether in Sydney or Brisbane, and takes over an exhibition centre, as it did in its first year at Sydney's Olympic site, and at South Bank Convention Centre, Brisbane. In subsequent years in Sydney it moved westwards to the demographic centre of the city, Rosehill Gardens Racecourse. Attendances on Fridays were invariably lower than at weekends. The Sydney Expo in 2008 sought to remedy this by having a Careers Friday that would encourage high school students to attend and learn about rural and regional employment options. Since Friday was

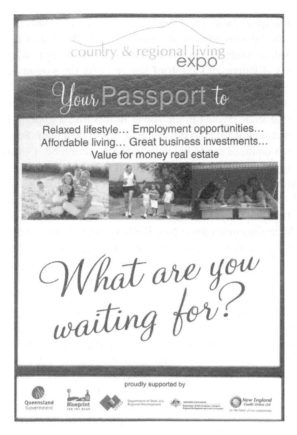

Figure 3.1   Passport for NSW Country Week Expo 2009
Source: Image used courtesy of Country Week Pty Ltd.

a school day the initiative failed. By 2009 the CW simply opened at midday on Friday and concentrated on the two weekend days. At all Expos local government councils, government departments and businesses were the principle exhibitors, paying to secure an exhibition space. As its 2008 website observed,

> Country Week, the Tree and Sea Change Expo, is your opportunity to find out about the virtues of living and working in country and regional Queensland or New South Wales. Country and regional areas offer professional and skilled trades people, as well as businesses, a wide variety of opportunities. Country Week is an easy, direct way to find out what is on offer.

Councils took up stalls of variable sizes according to their willingness and ability to commit resources to the Expo. Characteristically the larger towns, such as Armidale, Tamworth and Toowoomba, took up larger areas. Smaller towns, such as Warialda, Boorowa, Julia Creek and Thargomindah usually took up smaller spaces. Some small places combined resources and attended as a region. Eleven of the most distant councils in NSW, centred on Wagga Wagga, combined themselves into C-Change, or Country Change (with the slogan 'see the change, live the difference'). At some Expos stalls were set out geographically, so that those interested in, for example, northern Queensland could go directly to that part of the hall. While such layouts made sense in meeting a general feeling among many visitors that they wanted to move north, south or inland, they were difficult to organise when councils made belated decisions to attend. Consequently stalls were usually randomly distributed with council stalls mixed with others.

Visitor admission was free. Visitors were registered, and their names and jobs recorded, so that stallholders had an idea of what kind of employment or retirement strategies to focus on in discussions. Not all visitors took this seriously; the notorious Australian bushranger Ned Kelly wandered through the 2008 Sydney Expo. From 2009 visitors were issued with a 'passport' (Figure 3.1) that they could get stamped at the stalls that most interested them, and enabled those councils to follow them up.

For six hours a day council officials spruiked their wares, handing out literature and talking to visitors. Many dressed for the occasion in matching clothes that gave them and their own stall, itself given a local theme, presence and distinctiveness. Workers often dressed to emphasise distinct local themes. In 2007 Cootamundra's staff dressed as one day cricketers (distinguishing their town as the birthplace of cricketing legend Don Bradman). Dubbo came in red, Tamworth in green, Moree had long been in bright yellow, until 2008 when they developed an Art Deco theme, and were dressed as participants in a 1930s soirée, in keeping with the branding of Moree as a place with interesting Art Deco buildings, pictures of which adorned the stall. Posters and banners depicted Moree as the Artesian spa capital of Australia. Its staff promoted it as being twinned with a Czech spa town, Jesenik, while 'we also wanted people to say "Have you seen Moree?"', and made their stall so that 'it was like a home – homely – that you would want to come into and relax'.

Many had tee-shirts that conveyed their own slogans, and others were generic (such as Tumut's 'Family, Community, Lifestyle, Jobs' or Cabonne's 'Own Your Tomorrow'). Some were available for sale. Banners too tended to be generic; Temora was 'vibrant, safe and friendly', Young unremarkably suggested 'Live Young', though Yass took a different tone with a showbag emblazoned 'A Great Piece of Yass'. Boorowa too offered 'Superb Parrot, Superb Country'. Alongside the showbags and the decorated stalls (Figures 3.2 and 3.3) almost every place had sought distinctiveness to brand itself in a memorable way.

Parkes, a town that has gained fame through its annual Elvis Presley Festival (Brennan-Horley, Gibson and Connell 2007, Connell and Gibson 2011), always both promoted the Festival and had a life size, gold lamé-suited cardboard Elvis, a greater drawcard to their stall than information on jobs or housing (Figure 3.3). Glen Innes used their distinctive tartan in no small measure, reflecting their promotion

Figure 3.2   Oberon stall Country Week Expo 2010

Figure 3.3   Parkes and Elvis Country Week Expo 2010

of the town as the centre of Celtic country (see below). Condobolin had a CD of local pop star Shannon Noll (runner-up in the first Australian Idol contest in 2003) playing in the background, and copies were available for sale. Torres Strait dancers performed at regular intervals close to their council stall (Figure 3.4). Many stalls had local produce and other goods for sale or to be given away. Moree brought bottled water and pecan pie, Inverell had honeycomb (which turned out to be made in coastal Coffs Harbour), Muswellbrook olives, Narrabri jojoba oil and Oberon hydroponic tomatoes (Figure 3.2). Tumut had boxes of apples. Quirindi gave away seeds and Wagga Wagga native Australian trees. Queensland stalls displayed a cornucopia, from 'cool climate produce' from the south – apples and wines – to peanuts, bananas and mandarins from further north. Product marketing could be quite successful: 'I didn't know that so much produce came from these areas', while from Moree 'People are surprised to hear that we have all this produce'. Visitors tended to remember the pecan tart from Moree, Batlow apples and the Innisfail bananas (handed out by a 'cassowary') even if other details might have escaped them, giving those towns a profile and a small head start towards wider recognition.

Gunnedah, one of the most mining-oriented towns at the Sydney Expo, had a stall that included a huge lump of coal and, like some other councils, had dual clocks, designed to indicate both trivial commuting times and the possibility of

Figure 3.4   Torres Strait dancers Country Week 2008

going home for lunch (Figure 3.5). Glen Innes participants, on a nearby stall, were unimpressed with the coal symbolism; 'we are not sure it really helps to support the message we are trying to convey' of a rather different rural scene. By 2010 the coal had gone.

Some stalls mounted powerpoint presentations, or constant DVD films; others were content with posters, handouts and personal discussions. Manning Valley and Mackay gave away sophisticated DVDs as did some hospital boards, seeking nurses and other health workers. Many stalls had competitions, some designed for children ('Name the friendly bull'), that would keep visitors there a little longer. Some had particular promotions – 'win a weekend in …' – sponsored raffles for country produce and, in Queensland, offered prizes of rodeo tickets.

A tradition of most Australian shows is the showbag: a hessian, paper or plastic carrier bag given away (but purchased at more commercial shows) that contains such things as pens (embossed with 'Live in Canberra' and similar motifs), baseball caps, fridge magnets, suitably printed post-it notes, key rings, mints (and

Figure 3.5   Gunnedah's clock display Country Week 2008

GoWest chocolates) and even tea bags. The bags had their own slogans (such as 'Muswellbrook. Work, Live, Play', 'Sydney. Gateway to Lismore' and 'Discover Modern Country Living @ Dalby'). Perhaps the smallest exhibitor at any Expo, the tiny northern Queensland township of Thargomindah, with a population of about 250, proclaimed 'London, Paris, Thargomindah'. Some stalls added particularly distinctive items of their own; Cootamundra gave away rubber cricket balls, Toowoomba had tiny bottles of 'clean, clear crisp Toowoomba mountain air' and other Queensland councils offered sunscreen. The role of the show bags was to distribute information on the particular regions, usually some combination of a special edition of the local newspaper, tourist brochures, information for new residents, detailed information on house and land prices and sometimes jobs and festivals. The special editions of newspapers served diverse purposes. By informing local residents what was happening in Sydney and Brisbane on Expo weekend, they aimed to generate a sense of town and regional pride among existing residents, while offering a sense of place to visitors. Leaflets promoted festivals to encourage attendance, and in the hope that a visit to the town for a fun occasion might lead to a more serious intent to stay longer. At the very least a boost to tourism was a potentially valuable outcome of the Expos. The immediate outcome was that many visitors emerged with handfuls of showbags, having taken what was thrust upon them and preferring to take literature away rather than engage in detailed discussions with stallholders when they were uncertain about moving or had only vague notions of where particular places might be. Indeed, those with fewer showbags at the exit were often the more focused visitors.

Stalls were staffed by council workers – usually from economic development or tourism departments – but often included local estate agents. Councillors and mayors regularly attended, based on the correct assumption that visitors were impressed with being able to talk to a mayor – an unlikely situation in large urban areas – as some measure of accessible community. Other workers were simply private citizens with an enthusiasm for their towns. Sometimes these included people who had previously attended as a visitor, relocated, and returned as part of the local government-led delegation. At the 2006 Expo Oberon had as many as 16 local residents who had volunteered to help, but such a large number of volunteers was rare. While stalls may have had diverse participants, most were the respectable middle aged.

Alongside the more formal transfer of information via discussion and showbags, the Expo was also a show. At intervals of an hour or so demonstrations were held of 'country activities' such as whip cracking, while police from the Department of Corrective Services provided demonstrations of the ability of their dogs to respond and 'catch criminals'. At similar intervals seminars were given in a theatrette on more formal themes, such as financial provision for movement or the experiences of previous migrants. In 2006 a talk was repeatedly given by a Warialda resident, a quadriplegic confined to a wheelchair, who explained how a satisfying life could be had even with disabilities that might seem to deter rural residence. Raffles and land auctions sometimes also took centre stage.

Table 3.1   Exhibitors at the 2008 NSW Country Week Expo

Armidale Dumaresq Council	Moree Plains Shire Council
Australian Alpaca Association	Muswellbrook Shire Council
Birubi Sands	Narrabri Shire Council
Blowes Trousers	NSW Department of Health
Broken Hill City Council	NSW Rural Doctors Network
Business Enterprise Centre	NSW Department of State and
C Change Bureau	Regional Development
Coffs Harbour City Council	NSW Rural Fire Service
Cooma-Monaro Shire Council	Oberon Council
Cootamundra Shire Council	Office of Fair Trading
Countrylink	Onesteel
Cowra Shire Council	Pacific Dunes
Crighton Resorts Pty Ltd	Parkes Shire Council
Department of Ageing Disability and Home	QANTASLINK
Care	Raine and Horne Pty Ltd
Department of Community Services	Royal Far West
Department of Corrective Services	Stralia Web
Department of Immigration and Citizenship	TAFE NSW
Department of Lands	Tamworth Regional Council
Department of Primary Industries	Taronga Western Plains Zoo
Dubbo Visitors Centre and Events Bureau	TEACH NSW
Forbes Shire Council	Telco Asset Management Australia
Glen Innes Severn Council	Pty Ltd
Gowest	Temora Shire Council
Greater Taree City Council	Tumut Shire Council
Gunnedah Shire Council	University of New England
Gwydir Shire Council	Upper Lachlan Shire Council
Hunter New England Health	Weddin Shire Council
Inflight Magazine	Westpac Banking Corporation
Inverell Shire Council	Yass Valley Council
Kempsey Shire – Macleay ValleyL J Hooker	
Lachlan Shire Council	
Live in Canberra	
Liverpool Plains Shire Council	

Source: Country Week website, http://www.countryweek.com.au/news-media.htm#who (accessed 28 November 2009).

Other Participants

A range of other organisations also attended the Expos as at NSW in 2008 (Table 3.1). Most were state government departments, such as the Departments of Education and Corrective Services, anxious to encourage people to work for them in the regions. The Hunter New England division of NSW [Department of] Health, for example, brought generalised information on job opportunities and a special CW version of their newspaper (*HNE Health Matters*). The Rural Doctors Network and several hospital boards had stalls and similar missions for health workers. Queensland had many more public sector employers and a whole area separated off for the state government, and their elegantly red-shirted employees, reflecting the greater number of public sector jobs available in rural Queensland. Although many had relatively large, well-stocked stalls, they generated less interest than most council stalls. While teachers and nurses certainly came to the Expos, attracting new recruits into government departments proved challenging, because most visitors were uninterested in the public sector, and because years of training were usually required. A further disadvantage was that some Department stalls were staffed by metropolitan employees who were unfamiliar with regional circumstances in particular places.

A small number of private sector employers were also at the Expos. Only one was at the Sydney Expo in 2008 and otherwise there were none between 2007 and 2010. The lone participant, One Steel, a company employing 11,500 people and formed in 2000 when Broken Hill Proprietary Ltd. (BHP) privatised its steel division, displayed a large map of Australia with more than a hundred places with job vacancies, and generated considerable interest. By contrast the Queensland Expo in 2008 had a number of private sector employers, especially in areas related to mining, reflecting the greater number of companies working in areas where the economy was growing, and the need for workers. The engineering company, GHD, were aiming at gaining recruits but, if nothing else, at stimulating 'brand awareness'.

A third group of stallholders were representatives of housing estate development companies, some of whom were constructing residential development facilities and were at the Expos to sell these and retirement properties. Other organisations have also taken up stalls. The Country Women's Association (CWA), an iconic organisation formed in NSW in 1922 to represent the concerns of women and children living in the country, has attended to emphasise the social aspects of country life and encourage migration. 'Families are leaving the area and we haven't got the young ones out there any more' (Lesley Young, President, Country Women's Association, quoted in O'Dwyer 2008). In some respects, despite its own internal changes, the CWA was a residual of a rural lifestyle that smacked of embroidery, flower arrangement, jam and scones, rather than commerce and employment. It had no presence in Queensland and by 2008 it had stopped attending the Sydney Expo.

While CW has emphasised that the Expos are not about agriculture and should therefore be seen as distinct from the annual Royal Easter Show, when farmers come to the city for a two-week celebration of agricultural production, animals have intermittently appeared. At the 2008 NSW Expo animals returned for the first (and last) time in several years through a stall on 'Farming Alpacas' – designed to attract those who wished to make a distinct career change. A second stall – Blowes Trousers – a rural clothing store based in Dubbo and Tamworth – sold 'rural' hardwearing clothes, check shirts and moleskin trousers, alongside iconic RMWilliams boots. The clothes were not cheap, with boots being 'on sale' at $295, and Blowes did not do particularly well, failing to return in 2009. Ironically such clothes are either worn by a small number of more elite rural farmers and other residents, or by a conservative middle class urban elite, rather than the tradespeople who tended to be the bulk of Expo visitors. Few visitors sought agricultural life or its symbolic accoutrements.

Finally a handful of other private sector organisations have attended, usually intermittently. At the 2006 Expo yourseachange.com, a Brisbane coaching firm, appeared, and at the 2009 and 2010 NSW Expos, the consultancy group P2R (Possibility to Reality) offered 'structured and personalised coaching programs [to] help individuals, couples and families to make their sea or tree change', that centred on balancing options and alternatives, identifying and managing risks and being clear on what is important. Their promotional leaflet argued that

> Demographers estimate that 350,000 Australians per year think and talk about making a sea or tree change. 20 percent will make it happen. Of the 70,000 who make a sea or tree change, 20 percent have major regrets and give it away. 80 percent of sea or tree changers are glad they did it and only wish they'd done it sooner. Which category do you fit into?

At the 2009 NSW Expo a distinctive presence was rentafarmhouse, an organisation which had begun at Cumnock in central NSW in 2008, a township of about 295 people, some 50 kilometres from Orange and on a minor rural road. Neither Orange, which has experienced steady growth over recent decades, nor the surrounding Cabonne council area, usually attended CW, although outlying parts of the shire, like Cumnock, had experienced significant decline. When it was no longer economically feasible to support a school bus, and it was approaching ghost town status, a local resident instigated a scheme to rent empty farmhouses nearby, at $1 a week for up to three years, to people willing to move there, renovate the farmhouses and contribute to local life. In many respects it was a mini-CW but it demanded tradesmen who could renovate and were willing to live in some isolation.

Overall therefore a host of organisations with some claim to regional ties and regional expertise turned up, seeking people and especially workers. Numerically the Expos were consistently dominated by the repeated presence of various regional councils and those government departments with significant rural employment needs, particularly health, education and corrective services, where skills were required and jobs demanding.

Into the Marketplace

Local government councils have been the mainstay of CW since its inception. While in NSW this was usually a single council, or sometimes a combined effort by smaller local governments to promote a region, in Queensland the focus was on regional groups. Councils often initially attended to see if the investment in time and money was worthwhile, while internal politics and financial considerations were significant factors in their return.

Grouping of councils allowed human and financial resources to be pooled and a larger display constructed. Some councils saw this as beneficial in allowing an area to appear to offer more facilities and opportunities, a situation particularly true of smaller more remote councils: 'What one shire doesn't have the other shire does. The region is unique and diverse – that is what people want, some choice' (Tara Shire 2007). This also made it easier for some places to be recognised: again, for Tara Shire, 'Unlike Dalby or Chinchilla people don't know where we are located. Geographic location is very important. In that way it's better to market ourselves as a region' (quoted in Tsioutis 2007: 23). More frequently larger councils especially, such as Warwick, sought their own distinct identity: 'A collaborative effort was offered. But we deliberately said no, because we thought it would just dilute our identity. We are big enough to be by ourselves' (ibid). Perhaps unlike Tara, some smaller towns did lose their identity when the main representatives on stalls were from larger central places. In large amalgamated councils, such as Armidale-Dumaresq, small towns such as Uralla and Walcha are dwarfed by Armidale and, to their chagrin, tend to be downplayed in an emphasis on the larger centre. Smaller declining towns like Cumnock, in growing regions, had no opportunity to make a case.

Participation can be expensive: hiring a stall (a basic stall cost about $8,000), designing and printing material, transporting goods and people to the capital (often by air), hotel and food costs, overtime payments, and even cooking cakes. To be really effective, the costs also included producing a special newspaper, preparation work (including working with local businesses, notably estate and travel agents and also schools, to identify needs and opportunities) and follow-up work (contacting visitors, organising open days and visits). Larger councils and some parts of the public sector could absorb such costs more easily, whereas smaller and more impoverished councils, that might perhaps have most benefited, could not afford to attend. In Queensland councils were given state subsidies to attend. In NSW Kempsey Shire Council estimated a cost of $1,500 per person for 'travel and accommodation and sustenance for three days' (Kempsey Shire Council 2009), and most councils brought at least four or five workers. Several councils suggested that a round figure of about $16,000, not including the regular paid labour of council employees and the voluntary participation of elected officials and other community members, was the cost of participation. Alternative events, financial constraints and the inability to engage in everything, required choices. A factor that influenced attendance in 2006 was the emergence of the Inland Cities Alliance, with aims similar to those of CW but focused on larger regional centres, and reduced their attendance at the Expos. But it was short-lived and the larger centres – notably Dubbo and Albury – eventually returned. Similarly the Queensland government put on its own State Expo at the Sydney Olympic site, barely five kilometres away and on the same days as the NSW Expo. Competition and fission of this kind weakened the significance and influence of CW.

Councils' participation thus varied over time and space, being reviewed and debated annually by most councils. Commonly 'we could see value in it because it collectively gave us a voice in the city. It was one way for us to leverage some advertising cheaply and effectively'. But many councils that needed to promote themselves as destinations were not present at CW:

> It's the ones who can afford to do it. My only negative about Country Week is that it is run by private enterprise because those people who really need to be here ... they're not here because they don't have a budget for these sort of things. The communities that you know are struggling and really need to be here, they're just not here because the money's not there. They cannot afford to do it.

The perceived solution was:

> I'd like to see every local government in rural and regional NSW be here. Now that might be a perfect world scenario, but I think it would make for a much stronger Expo and the fact that you've even got to make the decision to come here and weigh it up economically – it shouldn't be like that. It's Country Week and everybody from the country should be coming down to the city. I think the public in Sydney would get more excited about that.

For a few councils the possibility of enhancing the conjuncture of a growing population and a growing economy was a central influence. Developing one would surely encourage the other. Towns like Tamworth, with an urban population of 42,000 and a further 14,000 in the local government area, and an annual population growth rate of about 1 percent, attended the Expos primarily to build on existing economic growth. Thus in 2009 the Manager of Economic Development stated:

> We're going down to offer an opportunity to Sydney people to change their lifestyle. Come to regional NSW, especially the Tamworth region, and discover a new lifestyle, safe and affordable living, and discover the endless opportunities that the Tamworth region has to offer in terms of employment, business and commercial opportunities. Tamworth has weathered the GFC very well. We didn't experience the downturn that the rest of the state has experienced. We're looking very, very good and we're screaming out for all sorts of employment, all sorts of people so we can grow our businesses. We're trying to attract the young families, the professionals, the tradespeople, and of course anybody who thinks they'd like a life change. Come to our town and help grow our own community (*Daily Telegraph*, 31 July 2009).

Employment deficits are the hard edge to softer evocations to lifestyle change, and that edge has hardened further as councils have increasingly targeted workers and families rather than retirees. Faster growing urban centres in the mining areas of Central Queensland were particularly anxious to acquire new workers, as their sole rationale for being at the Expos, dwarfing any significance of lifestyle changes, other than as the outcome of significant material gains. Such councils as Mt Isa, Emerald and Gunnedah were among the very few who were more concerned to attract workers than families; short-term needs were critical.

A number of councils sought to attract particular types of workers to meet the need of a resource boom, usually in mining:

> Certainly our reasons for being down here are very much focused on job seekers – trying to bring people in for employment. We have a list of about 45 jobs currently going. We have a wide range of jobs from pruning and harvesting, menial labour, through to a number of graduate positions.

> It's skill shortage. That's the issue. We're trying to attract people who are living in Sydney. Our research indicates that 25 percent of the Sydney population is thinking about moving to inland Australia, so tree-change rather than sea-change. We're trying to get to those ones and present our region as a viable option to them. We've actually got real jobs that we have brought down where we've got gaps, and particularly in some of the more professional areas such as accountants; we're desperate for accountants, we've got a brief for a stockbroker, and all the trades – we need one of everything.

But in many respects the rationale for CW is that most towns are not growing at all, let alone as fast as the Queensland mining towns, so that even a handful of new workers and businesses would be more than welcome. Simply being at the Expo in hope was therefore deemed important. An exhibitor from a relatively small town was sceptical about the benefits:

> We miss out to the big, nearby places so it's hard to say what we got out of it last year – just tradespeople who knew we were here and came to talk to us – but we have to be here – we have to have our image in the marketplace.

Not all councils, presumably including those who dropped out, accepted this need to be in the marketplace at some cost. Nonetheless several participated simply because they did not want to be left out, believing that they might somehow be disadvantaged by not having a presence. More frequently councils had developed specific goals, notably to attract residents, expand existing businesses and services and secure economic growth, and the marketing to achieve them. With population decline, aging populations with limited spending power, or reduced demand for services such as schools, there was concern that the school might lose a teacher, or the town a police officer. Some of these people would be community leaders, with a community role that was then lost to the town:

> The specific market we target is families, young people and families. Most rural and regional communities are losing that 18 to 40 year old demographic. They

just go and don't come back. Any one that we can have with young families is welcomed with open arms because they provide for teachers, doctors, chemists and all those professionals that require families to survive.

Some councils recognised that they were in a stronger position than those that were predominantly agriculturally based:

Our advantage comes from the fact that this town has had substantial investment in the timber industry, particularly in value-adding ... so we have had alternative employment options other than agriculture.

Councils usually continued to attend the Expos because they had genuinely interested visitors. Several had learned the difference between the number of enquiries and the importance of fewer, but more genuinely interested, enquiries:

It's a lot quieter this year, but the inquiries are a lot more solid. There's less tyre-kickers and more people who are genuine. We've taken people to the computer and shown them that there are three positions listed here, and we know there are more.

The way we saw it, it is the only opportunity in Sydney for people interested in coming to the country ... We went to the Royal Easter Show three years ago ... and people weren't in that mindset. All they were interested in were showbags and rides. What we discovered was that there was interest in moving out, but on that day, in that particular setting, people were not interested whatsoever ... you have to attract them at a place where their mind is thinking about it ... Country Week is the only dedicated show where people are in that mindset.

Effective participation at CW could help councils promote development, fill employment vacancies and retain services. One council recognised that it also performed a positive role internal to the council, since it helped to cross boundaries within the council structure, gave people the opportunity to work together and solve problems, and present a positive face to outsiders: a process analogous to that described in central Sweden as 'internal marketing as identity formation' (Cassel 2008). Improved communications created optimism:

I think the biggest benefit is that confidence is growing in the town because we are promoting ourselves. You start to get proud of who you are, your identity ... It also helps us with team building as an organisation and it really cemented the relationship between the new Mayor and the executive team.

Some councils saw the Expo as one of various possible activities, and incorporated decisions about attendance into an ongoing cycle of participation and review. Where the Expo was seen as an event which required stand-alone justification and hard evidence of success, councils were unlikely to return. Conversely councils that saw CW as part of a process of working with the local community and media, conducting follow-up activities, realising there would be a lag time in achieving results, and recognising that it was difficult to measure the impact of participation, were likely to remain. Multiple reasons accounted for participating in CW, not least the personalities of dominant councillors and public servants, and for diverse expectations about what it might achieve, but all were centred on increasing populations to meet employment needs or prevent decline. However a more comprehensive and effective evaluation of what the benefits of being 'in the marketplace' actually were, and how a more positive message about these benefits could be diffused, remained elusive for most councils. Few had the methodology or capacity to monitor the impact of the Expos.

A Geography of Participation

Not all councils participated in the Expos, and the number of councils attending has fallen slightly since 2005. Some attended once, but never returned. While these councils were not interviewed directly, others made comments about their expectations and approach. Most saw their expectations as being too short term, or their not having used the Expo effectively. One council that had been there for three years noted;

> the electronic collection of information [is crucial] ... What we found was that the two adjacent shires were not swiping any of the cards. They were good at shaking hands, they were good at talking with people, but they were not good at gathering information and if you don't have that point of contact you don't have the ability to follow up.

The social and technical inability to follow up queries and visitor interest necessarily resulted in poor outcomes and a disinclination to return. However geography was a particular influence on participation.

Coastal councils such as Coffs Harbour were less likely to attend the Expos both because they had little trouble attracting people to that part of the coast (especially as the baby boom generation reaches retirement age) and faced some degree of congestion, and their own aging population. However, rapidly expanding Gold Coast attended the 2008 Queensland Expo with the specific goal of attracting skilled workers (rather than retirees) that are essential for effective management of a growing city. Employment needs outweighed environmental concerns. Similarly Coffs Harbour/Bellingen returned in 2008 after a gap of three years, but was absent again thereafter. Despite growing relatively rapidly they sought particular kinds of workers – a new concrete plant was being established and 120 new workers were needed – while two new retirement homes had recently been constructed, hence they were not uninterested in retirees. They also wanted to get their own children back: 'we know that our children will tend to leave at 18 but we want to get them back when they are a few years older and starting their own families'. Those who had previously come from the regions were seen as having the greatest potential to fit in and stimulate appropriate local development (see Gabriel 2002). The only coastal region that has routinely come to the Expos have been Nambucca and the Macleay Valley, centred on the town of Kempsey, midway on the NSW coast. Both Nambucca and Kempsey have experienced considerable economic growth, and so seek a labour force for small-scale manufacturing industry, and offer numerous attractions for sea change. What has mainly slowed growth has been their distance from Sydney (and Newcastle). Advertising Kempsey was not particularly challenging, as Peter Bailey observed in a newspaper Expo promotion:

> They're on a major highway, they've got air links, they're a community that could easily potentially grow quite significantly if they can get some better exposure which is why they come down to our event'. With a population of about 30,000 the Macleay Valley Coast boasts 80 kilometres of pristine coastline, wild national parks ... top medical facilities, diverse employment opportunities ... Several progressive businesses and industries have already accepted the invitation ... To the sea- and tree-changer lifestyle is paramount, and Kempsey's range of outdoor activities offers everything from swimming to surfing, fishing, sailing, kayaking, windsurfing, water-skiing, scuba diving, bushwalking, cycling, tennis, golf and bowls. Culture and nature buffs can explore art galleries, a local Arts Trail, historic walks and rides, heritage villages and wineries. Sydney's foodie revolution has also extended its reach, with Kempsey's restaurants and cafes serving fresh local sea food, beef and produce – and save room for a generous helping of Kempsey's famous country hospitality. The area's pub and clubs also provide regular social activities, plus the essential agreement for toasting your successful escape from the grind: ice-cold beer (*Daily Telegraph*, 31 July 2009).

Such notions of lifestyle, centred on leisure, routinely pervade CW's literature and bring some promotional material close to tourist brochures, but especially in coastal areas. The coastal Macleay Valley, with 30,000 people and a solid economy, alongside Nambucca, presented fewer marketing challenges than small, remote inland centres.

Some inland councils have also dropped out. Between 2005 and 2006 most of the councils from far western NSW (such as Wagga Wagga, Deniliquin and Bourke) dropped out as did several from the Hunter Valley, though they later returned. The councils that regularly attended thus represent a broad band through central NSW, excluding the far west and the coast (Figure 3.6). Similarly in Queensland more remote northern councils were least likely to attend, though they came from the far west. As the mayor of the small western town of Quilpie pointed out: 'Some western shires aren't here probably because they are so different, as they really are bush bush – very different to the city. It might be a tough crowd for them here. We're different because we have water' (quoted in Tsioutis 2007: 28). Most

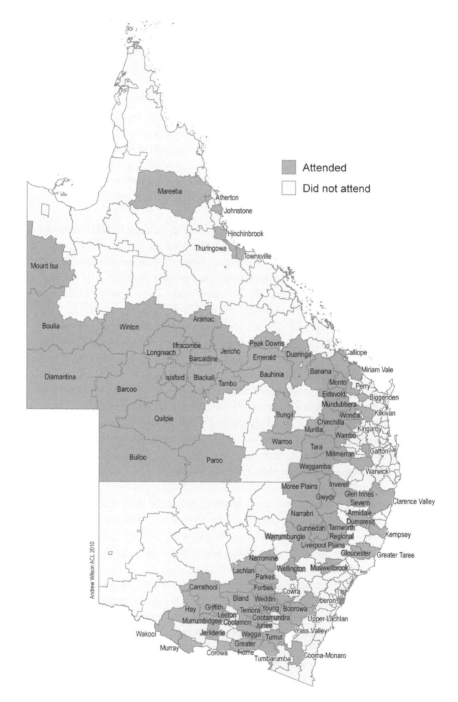

Figure 3.6   NSW and Queensland Councils. Country Week 2007

dropped out because of limited evident success from previous Expos, notably smaller councils for whom the expense was significant and the results did not justify participation. Uralla Shire decided to attend only every second year because of 'limited tangible interest' at the 2004 CW (but was then amalgamated with Armidale), and Upper Hunter, Coonamble, Warialda (Gwydir) and Tenterfield also withdrew after unsuccessful experiences. Moree Plains, one of the most remote councils to attend NSW Expo, after years of being one of the most innovative (and twice winning the annual award for the best stall), failed to return in 2009 after a new council decided the returns were inadequate, but reappeared in 2010.

Rather like coastal councils, nearby inland councils not far from Sydney, such as the Southern Highlands (which includes Kangaroo Valley, a popular Sydney second-home area), did not usually participate, though the Southern Highlands briefly broke that tradition in 2009. They were there 'to see how it works' and because 'we need workers of all kinds, including labourers but especially skilled workers. Many commute from the Illawarra [Wollongong] but we tell them to come and live here'. Indicative of the reason why they had hitherto not attended were their selling points: 'We have everything here – country living within reach of the ocean. It's just a two hours drive to Macquarie Street [central Sydney] without a single red light. It's a joy to come home to at night'. They scorned some of the efforts of other councils with displays of wine and food that they felt seemed to trivialise the event. Larger towns in NSW, such 'sponge cities' as Orange and Bathurst, did not normally attend since their populations have generally grown and jobs have been partly filled by commuting from nearby smaller towns where jobs have been lost. Nonetheless Tamworth and Armidale were regular participants, and Wagga Wagga usually came. The national capital, the rather more wealthy Canberra, always took a large stall, and many large Queensland cities were there because of their demand for workers though all experienced steady population growth.

Councils came to the Expos to stimulate increased interest in their region, minimally to put themselves 'on the map', and hopefully encourage some migration, or perhaps tourism. It was not a venue in which to encourage investment, and councils that expected new business migration were usually disappointed. One of many ironies of CW is that those councils most anxious to encourage migration and where populations are actually falling (e.g. Lachlan and Moree Plains) were those that were least familiar to urban residents and most distant from them, hence experienced the greatest challenges to successful participation. Marketing quite small towns, largely unknown to metropolitan residents, constitutes a considerable challenge that not all councils could meet. How councils developed strategies, and marketing and branding themes, can now be examined.

Chapter 4
Strategies: 'In It to Win It'

Some of us just don't know how to market ourselves. We have to debate things
like – is 'quiet' a good thing? (Economic Development Officer, Grenfell, 2007)

Place marketing is the core of Country Week (CW), hence this chapter focuses on
the branding of place by local councils, in a context of considerable competition.
It gives particular attention to Glen Innes and Oberon, where recent migration
is later examined in greater detail, and the very different town of Gunnedah, to
reflect on the similarities and differences in marketing and mobility. Marketing
and branding take multiple forms, from the decoration of stalls and the language
used to promote places, to the council fliers, tourism brochures and the special
editions of newspapers. Some stalls had other forms of publicity, from DVDs to
beer mats and car windscreen stickers. Not all can be reviewed here, but examples
highlight both typical and unusual approaches to place marketing, though rarely is
it possible to indicate what has proved particularly successful. Even the most basic
logo, that of CW itself, along with the name of the organisation and of the Expos,
has varied over time.

Councils (and other organisations) were both competing for visitor interest and
competing against each other. They therefore needed to distinguish themselves,
by strengthening their own image and attributes. Some councils relied on cultural
heritage, such as regional Scottish or Irish ancestry, while others used the physical
environment. Boorowa featured sheep and parrots, and Moree emphasised its natural
mineral spas. Idyllic locations often set the scene: 'People are firstly drawn by our
photos of green and lush rainforests, and then we give them our information'. Every
council saw branding as a learning process: 'We've learned with Country Week the
notion of branding and other things about our area. You'll see it with the colour of
our shirts, the logo'. Some branding was initially based on lifestyle attractions, or
resembled tourism promotions, but most councils quickly recognised that lifestyle
was not the key motivation for visitors contemplating moving. Branding became
more closely linked to generic employment opportunities, housing, education and
health facilities, but alongside particular local features. Although Elvis Presley
might have attracted visitors to the Parkes stall, he could not keep them there. As
one council Economic Development Officer stated:

We didn't want to promote 'live it, love it, do it' which is very generic. It's no
good just coming in and having a good time and just handing out brochures. One
of the things that people liked on our stand was the roses. Warwick is regarded as
the 'rose capital' of Queensland. But having the roses there is not going to make

the people move, it will just give them a bit more brand than just location – to remember them by ... and image association (quoted in Tsioutis (2007: 38-9).

Developing a branding theme that was both specific to a particular place, eye-catching (and ultimately accurate and informative) was rarely easy for councils with limited experience and expertise in marketing. Once again larger towns, perhaps with more to offer, and more expertise to draw on and pay for, tended to be more successful. Some councils maximised use of resources by preparing materials that could also be used elsewhere: 'We decided that we had to come because we had to have a presence ... and I decided for our involvement that I had to get take-outs that would last beyond Country Week, so this brochure will have a life of 12 to 18 months'. Portable and flexible, generic or specific, emphasising natural or human landscapes, promotions varied between places and over time.

Getting There

CW alone promoted the Expos. Promotional brochures were distributed to households in the perceived target areas. In Queensland in 2007 this included some 300,000 households in Brisbane and south-east Queensland, more or less the urban population of the region. In the same year brochures were distributed to 160,000 households in western and north-western Sydney. Rather than target the whole population of Sydney, the market was defined as the rather less prosperous parts of Sydney, where tradespeople and 'battlers' were more likely to live, housing stress was more probable and which were also relatively close to the venue. Brochures were supplemented with advertisements in newspapers, both regional free newspapers in western Sydney, and the Sydney *Daily Telegraph,* the newspaper more likely to be read by tradespeople, and by radio advertisements – once again on commercial radio stations more likely to be listened to by residents of western Sydney. Western Sydney is generally of lower socio-economic status than eastern Sydney, but house prices are high and rental accommodation is expensive and scarce relative to regional areas; it also experiences a high level of localised traffic congestion, considerable highway toll charges, and an inadequate public transport system. Most interest was stimulated by local radio stations running brief promotions or occasional interviews. Advertising was constantly a sore point for stallholders who never felt they had enough of the 'right' visitors, and for many visitors who complained they had only learned about it at the last minute.

Choosing the Targets

While Councils were principally interested in attracting households, particularly households that could make a wider economic contribution to the community, they were also interested in the migration of businesses. While some potential migrants had businesses, or were interested in changing career and developing a small business of their own (Chapter 6), stimulating the migration of larger businesses with significant workforces was particularly difficult and not the focus or outcome of the Expos.

Over time the target population gradually became young households with children. As Cootamundra pointed out: 'They'll send their kids to school, open up businesses, contribute to the community … pay taxes … We're looking for young families in order to grow' (quoted in Tsioutis 2007: 30). A much vaunted measure of success was tiny Thargomindah gaining a butcher with five children. However younger single people were not targeted despite their migration from regional Australia, since they were seen as unreliable and unlikely to persevere and remain. In Roma 'We've found that younger people don't want to work – they just get a job to save up money and then leave the country' (op cit: 34). More mature people, in every sense, were more likely to remain. However, while retirees were welcome, councils gradually became less interested in encouraging retirement migration. Although older people may purchase houses and commodities, and had more time to join local organisations, they were not key targets.

Most councils believed, often erroneously, that it was relatively easy to encourage the migration of the retired; as one CW committee member said: 'retirees and grey nomads: they are a walk in the park – but we need to get the battlers'. In some towns services were already overstretched, and residential accommodation for the elderly limited, evident in both Glen Innes and Oberon (Chapter 7), and most of regional Australia has a high older dependency ratio. As a Tumut council worker phrased it: 'The retirees can go somewhere else. We want workers'. By contrast residential estate developers saw them as a valuable market, though most residential estates were in coastal areas.

There were expressions of desperation, especially where populations were declining, but also where demand for workers was high. In Toowoomba 'we really need people; anyone will do' while, more strategically, Innisfail sought a 'slightly higher population as we need a bigger rate base' (quoted in Tsioutis 2007: 29). In inland regions everyone had the potential to make a valued contribution simply by being there. Bringing relatives was a bonus.

Marketing Glen Innes, Oberon and Gunnedah

Glen Innes has attended the NSW Expo in every year since it began, and had quickly become conscious of the need to differentiate itself from elsewhere. The Deputy Mayor noted that at CW 'people are bombarded with information … they need some sort of branding to separate Glen Innes from other towns. Our Celtic heritage is unbelievable for providing awareness of and a positive attitude towards Glen Innes' (Davidson 2008: 3). The use of that assumed Celtic heritage included tartan decorations of the stall, dressing some of the workers in Celtic attire, distributing flyers to promote the annual Australian Celtic Festival and even including tea bags with tartan tassels in their showbag. While branding Glen Innes as 'Celtic Country' has a tenuous link to local history it has been a deliberate and effective branding exercise that has also stimulated tourism (Connell and Rugendyke 2010).

Behind Celtic Country, and Minerama (a large gem, mineral and craft festival held in March each year), was a concerted effort to promote basic themes that visitors were usually looking for, including access to medical facilities, aged care facilities, education (including electronic access to the facilities of the University of New England, with its main campus in nearby Armidale), locational advantages for businesses and affordable housing. To what were largely generic themes Glen Innes also stressed the environment, described in their 2009 newspaper under the heading 'A great place to live, work and play' as including a 'high average annual rainfall and the beauty of four distinct seasons'.

Glen Innes was also constructed as 'a district on the move', with population growth being cited as evidence of progress, and an image of a newly constructed McDonalds Family Restaurant on the front page of the special edition of the 2008 *Glen Innes Examiner*. That perhaps debatable image of progress connected city and country – a form of suburbanisation, through the arrival of fast food in rural areas – in ways that both problematised notions of traditional rural living, and perhaps 'slow food', and emphasised the presence of formerly absent urban amenity. Like other places where the juxtaposition of images is debatable, there was no apparent conflict or disjuncture between images of the ubiquitous and modern McDonalds and the Celtic heritage. Glen Innes could be perceived, branded and marketed in various ways.

Oberon had also attended every Expo and for many years marketed itself as a 'family destination' with its key message for potential residents being 'relocate without losing touch'. Scenery was promoted as a particular feature of Oberon, under the title 'Oberon – Simply Spectacular':

> Moving to Oberon means you can enjoy all the country lifestyle has to offer – clean air, little traffic, a crisp alpine climate, a short drive or walk to work (which means less time commuting), friendly people, a real sense of community, a safer environment, and some of the most spectacular scenery in NSW.

As with many other towns, Oberon was positioned in the 'heart' of a region or route, rather than being a remote settlement, here as 'The Heart of the Tablelands Way', connecting Canberra to Muswellbrook. Through the use of concentric circles, Oberon was portrayed as being two hours from Mudgee to the north, Goulburn to the south and Sydney to the east (Figure 4.1), only one of which was likely to be of much interest to visitors or residents. As a small upland town – 'heaven on a hill' – its Mayor reluctantly noted: 'Of course we don't have everything – no McDonalds or KFC etc' (*Oberon Review*, 5 August 2010).

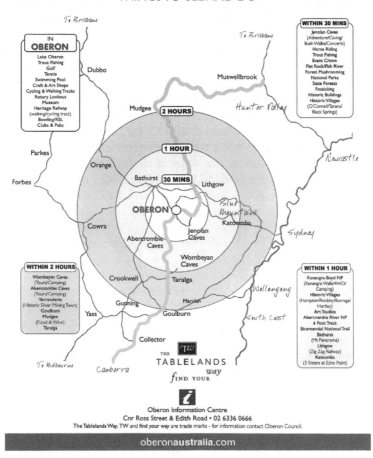

Figure 4.1   Oberon in place, Country Week 2008
Source: Image used courtesy of Oberon Council.

Oberon was marketed particularly to young families. The story of Scott Dawson and Jodie Barton, and their daughter Eliza, was prominent in the special 2006 issue of the *Oberon Review*, under the heading 'No pollution, no traffic, plenty of reasons to stay'. The family had attended the 2005 Expo, and moved from the western Sydney suburb of Blacktown to Oberon, with Jodie 'still working a few days a week back in Sydney'. Their successful transition was similar to that of

Dave Ward and Krysta Katz, in the same issue, who with their young daughter, had also moved from Parramatta in western Sydney, with Dave saying 'many people for some reason think Oberon is far removed from Sydney, really it's only a few hours away, so it's quite close'. Many promotions used similar local success stories modelled on those of CW itself and, as here, emphasising amenity and accessibility (see Chapter 8). By 2008 the brand message was 'a great place to visit … a better place to live'. As in similarly high-altitude Glen Innes, the 'distinctive climate' was marketed as a year-round advantage. In winter there were 'glorious layers of snow' (rare in Australia) and in summer 'mild summers are a real treat in Oberon, unlike the sweltering summers in the city'.

The natural features of Oberon were augmented by cultural activities, including activities that accentuated the natural attributes of the area, such as the Oberon Daffodil Festival of Spring Gardens which allowed residents and visitors to see the private gardens of Oberon residents. These were also highlighted in Oberon's promotion of outlying townships, the 'Oberon circle of villages', a rare example of a town promoting its hinterland. As in Glen Innes, with its Celtic Festival and Minerama, festivals were both a means of indicating that towns had potentially lively activities and offered tourism opportunities, while also celebrating distinctiveness and heritage.

Gunnedah, a rather different town, where coal mining and related activities grew in the 2000s, took a rather different perspective. It too was an annual participant at the NSW Expo, and over the years refined its message, from initially promoting tourism-related themes, as the 'Koala Capital of the World', to developing messages that promoted employment, housing and lifestyle opportunities. The main slogan became 'land of opportunity', and its two distinctive exhibits were a piece of coal and a series of clocks that indicated the daily lifestyle that time-poor urban dwellers could enjoy if they relocated (Figure 3.5). Gunnedah primarily sought to attract skilled workers to maintain the economic momentum of the mining boom. Leaflets highlighted that '$300,000 can purchase you a *three bedroom home* in leafy streets, close to schools, the swimming pool and hospital' (Gunnedah Shire Council 2009; original emphasis).

Gunnedah had marketing advantages that other councils could not claim. In addition to financial support from mining companies to promote the town, Gunnedah also had 'former Gunnedah girl-turned-international supermodel' Miranda Kerr to promote both the town and the plight of the koalas. According to the council's economic development officer, she 'approached the council with the idea of helping publicise the plight of the koala and, in the process, promoting the shire'. She was duly publicised in Gunnedah's brochures. Dead or alive, certain personalities, from Don Bradman and Elvis Presley, to Shannon Noll and Miranda Kerr, were drawn into the struggle for distinctiveness.

Gunnedah recognised that people would move to the town not because it was the 'Koala Capital of the World', but because it could offer potential residents employment, affordable housing, education and health services and a lifestyle that eliminated some problems of big city life. Proximity to Sydney made it relatively easy for people to visit Oberon, and council strategy was to promote festivals and encourage visits, so that perceptions of distance, strangeness and isolation would diminish. Glen Innes, 800 kilometres from Sydney, had little chance of encouraging casual visiting, but hoped for 'early retirees', often coming from Queensland. Otherwise two towns of similar size both focused on environment, community and, especially in Glen Innes, generic housing and employment issues. Oberon, close to Sydney, mentioned lifestyle more than most places. Each council stressed, as far as was feasible, a diversity of opportunities. Ultimately the marketing strategies of each of the three towns were remarkably similar, and as the focus on employment and housing grew, converged further.

Multiple Choice, Multiple Images

All councils produced brochures and information packages for the Expos. They varied in quantity and quality as a result of financial resources available to councils, production skills and the knowledge of what visitors might be seeking. That knowledge came from experience or was disseminated by CW. As one Queensland Economic Development officer said: 'Peter Bailey came around the whole region and told us how to do things. What people are after – health, education and real estate. And that we also need to have a good cross section of people on the stand ... community people ... the Mayor' (quoted in Tsioutis 2007: 49-50). Otherwise acquiring knowledge took time. A Queensland Mayor noted: 'People really are chasing land information. We should've thought of having agents or advertising at our stall. We should've had that information ...' (ibid). Nonetheless the many brochures and newspapers recorded how the towns perceived themselves, and how they would like to be seen by prospective new residents.

Basic branding, before any attention to detailed content, created slogans for immediate response. Inverell was 'The Sapphire City', Outback Queensland was 'LEGENDary', St. George the 'Window of the West', Hay 'Big Sky Country' and Narrabri 'Cotton Country'. At Julia Creek 'You feel free in the wide open spaces'. Narrabri was also the 'Heart of the North West', Lachlan 'The Heart of NSW' and Yass 'The Heart of Capital Country'. As at Oberon the heart hopefully implied the central part of the body (or region), an important organ (or town) and caring.

A widespread theme was the notion of 'country', which varied between places, and in many cases involved hybridity. Cootamundra professed 'new country living', never defined in brochures and information sheets that discussed transport access and real estate prices. Boorowa promoted their shire as 'Country with Character' and Tamworth offered 'City Style–Country Heart'. Tamworth's newspaper showed an image of well-dressed diners at a restaurant enjoying wine, coffee, pizza and each other's company. The food, coffee and wine evoked images of city restaurants, alongside community and belonging. Dalby assured visitors they would 'enjoy a country welcome' and encouraged them to 'Discover modern

Table 4.1   Themes present in selected rural newspapers available at Country Week Expo

Newspaper	Mayor's welcome	Stories of recent migrants	Jobs	Housing	Education	Medical facilities	Recreation	Local attractions
Eastern Riverina Chronicle (2009)	yes	yes	no, but yes for business opportunities	adverts	yes	yes	yes	yes
Oberon Review (2009)	yes	yes	yes	yes	yes	yes	yes	yes
Forbes Advocate (2008)	yes	no	no	yes	yes	yes	yes	yes
The Guyra Argus (2009)	yes, in a story	no	regular listings	regular listings	no	yes	no	no
Muswellbrook Chronicle (2006)	yes, in a story	no	yes	yes	yes	yes	yes	some
Yass Tribune (2008)	GM welcome	no, vox pops of residents	no	yes	no	no	yes	yes
Manning River Times (2009)	yes	no	no	yes	yes	yes	yes	yes
Boorowa News (2009)	yes	yes	no	regular listings	yes	yes	yes	yes
Crookwell Gazette (2008)	no	yes	no, but stories of work	yes	yes	yes	yes	yes
Parkes Champion-Post (2009)	yes	yes	yes, a little	yes	yes	no	yes	yes
Tumut and Adelong Times (2006)	yes	no	yes	yes	yes	yes	yes	some
Warwick Daily News (2008)	yes	yes	yes	yes	yes	yes	yes	yes

country living @ Dalby'. The country was no traditional agricultural concept with limited appeal to urban dwellers, but blended rural and urban benefits.

Councils regularly emphasised progress. Tamworth promised a 'bright and secure future', the region was 'riding high with equine centre (sic)' and 'Up, up and away' with a new shopping centre. Tamworth also produced headlines such as 'Looking to the Future' and 'Strategy to look beyond 2010' – versions of which were evident elsewhere. Many councils settled on their natural resources, including good agricultural soil and proximity to national parks. Notions of sustainable development occasionally surfaced; Tumut, which claimed to be 'Developing Naturally inc.', aspired to a convergence of nature and development in their Vision Statement: 'We are a (diverse) rural community, working to improve the well-being, welfare and prosperity of all people in our community, whilst caring for and sustaining our environment well into the future'. More frequently the natural environment was simply to be enjoyed.

Special editions of local newspapers took up similar themes, usually with a more personalised focus, through an emphasis on the experience of people who had moved, as in Oberon, a link to the local councils and dignitaries, and actual local life, as a random sample of such newspapers indicates (Table 4.1). Front page headlines usually had generic themes. The *Crookwell Gazette* proclaimed that 'life is local', in Muswellbrook you could 'work, live, play' while Forbes was 'friendly-historic-inviting'. Such slogans drew on geography, history and local political economy. Employment opportunities were less prominent, presumably because in some instances there were very few to promote. In 2010 neither Oberon nor Glen Innes had specific jobs available. Some publications, notably the *Crookwell Gazette*, carried stories of local people at work, projecting the town as a place where people could find employment though no specific job opportunities were promoted.

Content sometimes read like a checklist. In *The Forbes Advocate*, the council had a full page advertisement based on a jigsaw puzzle. Under the heading 'You will fit into Forbes', there were 12 jigsaw pieces, comprising of a location map, and then the following words accompanied by relevant images: real estate, work, play, education, health care, industry, transport, community, history, nature and events. Similarly the 2008 *Warwick Daily News*, which featured two couples with children and headings 'family leaves city life behind', 'Brisbanites relocated and discover benefits of life west of the Great Divide', 'City's best lured by tree change' and 'bring up their young family in a safe environment', filled the remaining 21 pages with sections that included: Wineries and Gourmet Produce, National Parks, Economic Boom, Local Employment, Excellence in Education, Medical Facilities, and Housing and Property.

A focus on local life could have drawbacks. More detailed reading of even the special editions revealed problems and disadvantages, in learning about swine flu (*The Guyra Argus*) or cancer (*Muswellbrook Chronicle*). Many regional newspapers conventionally included an agricultural supplement, and this may have both emphasised some of the challenges faced by agriculture (prices, droughts etc) and may have given some readers an unduly agricultural perspective of local activity. Occasionally accidents (literally) happened. In 2008 the Cooma council realised that the special edition of one of their two local newspapers had a banner headline 'Highway fatality' and became somewhat reluctant to distribute it. The front page of the newspaper of one Queensland town seeking more workers covered the closure of a major nearby mine. In 2010 Yass triumphed its 'excellent health care facilities' while the pages of the enclosed standard edition emphasised that 'the staffing crisis at Yass Hospital continues unabated'. Somewhat differently Thargomindah's stall was dominated by an elegant aerial photo of the township, which highlighted the red soil and dust. A front page headline from the *Parkes Champion-Post*, 'Two cars collide, no-one injured', might have suggested to some not so much an image of safety but one of a place where very little happened.

Councils sought to brand and market their towns in the most effective way, by focusing on four key themes: lifestyle (including community, scenery and tourism), employment, housing (and other services) and some distinction from metropolitan life (without spiralling house prices, congestion, violence or the lack of community). Most places linked all the key themes together in some way, in brochures, newspapers, DVDs or verbal sales pitches, which usually initially mentioned lifestyle followed by the 'realities' of employment and housing. Brochures were predictable in being comprehensive, suggesting a diversity of opportunities and a sense of progress but in a hybrid country where the best of urban and rural life came together.

Lifestyle and Scenery

The most established element of CW's logo has been 'Live the dream': a powerful and pervasive theme in all marketing, even broadcast over loudspeakers during Expos, but not without criticism. As a former CW official stated: 'Well, that might work in Mosman [an elite Sydney harbourside suburb] but there is a different reality elsewhere, a reality of congestion, tollway costs and lack of amenities'. Most stalls and publications offered slogans and banners with generic catch phrases such as 'Live the lifestyle', 'It's all about lifestyle' and 'The complete lifestyle package'. Photos of wide open spaces, people enjoying leisure and the availability of fresh produce supplemented the slogans. However over time there was growing recognition that lifestyle needed to be grounded in more tangible contexts.

At the core of CW's perception of country life, and 'the dream', was that migration enabled valuable lifestyle changes, and these themes were taken up by participating councils and by those who had moved (Chapter 8). Such changes were often quite amorphous despite frequent references to more family time, less travel and more recreation, fresh air, community and neighbourliness. The epitome of such a lifestyle change was a move beyond even small town life towards the adoption of country pursuits. In promoting the 2009 Sydney Expo Peter Bailey, claimed that increasing numbers of people were now saying:

'If we want to get out of town, we want to pursue the dream, we want the five acres, we want the pony, we want the great lifestyle' and so lots of people also make the decision that if they're going to move, that they're going to have a few acres on the edge often, and that's very achievable (quoted in *Daily Telegraph*, 31 July 2009).

Perhaps, more loosely, it was 'more exhausted parents and rundown workers opting for a golden trifecta: a more relaxed lifestyle, a sense of community, and affordable real estate' (*ibid*). Relaxation and 'family time' were recurrent themes. Like that of Moree, Coolamon's stand was set out as a lounge room, with coffee table, and the theme 'have a seat, relax, take a breath'.

While few people had distinctively rural objectives in terms of space (Chapter 6) a focus on rural lifestyle was important in counteracting subversive drought and 'dead sheep in the dam' perceptions. For several councils one of the basic reasons for being at the Expos was to spread the message that, while some places had experienced problems, others had not and regional Australia was much more than the sum of the problems. Goondiwindi sought to 'get the message out there, that rural areas are not all about negative images they see on TV'. At one level this was generic with all councils offering positive messages and, notably in Queensland, pointing out that access to water was often better in regional Australia than in Brisbane where water shortages were not unusual. 'We're green' was a recurrent message, and at Innisfail, 'water, basically unrestricted water ... I think that's a big thing. We are close to the wet tropics rainforest; we have no shortage'. Inland water security briefly ushered in notions of 'oasis change' (Dickins 2007).

CW also sought to emphasise the diversity of regional Australia and that different places might meet the needs of different people:

What we are trying to convey to people in Sydney is that country or regional NSW can be what you want it to be. If you want urban sophistication, theatre, comedy festivals, foreign film festivals and cultural diversity, we can take you to big cities like Newcastle, Tamworth, Armidale and Dubbo. You'll have a far better lifestyle, with all the services you have in Sydney. Medium-sized centres have what I call 'connected' cities where you walk in the main street and know people, or you could go to a smaller country community like Boorowa, down near Yass, and it's one of the friendliest communities I've ever been to, a great place (Peter Bailey, quoted in *Daily Telegraph*, 31 July 2009).

Once again the country might claim all the positive attributes of urban life. Ironically Newcastle had never been to CW and Dubbo rarely went.

At the earliest Expos councils tended to emphasise the touristic aspects of regional life, concentrating on physical amenity and landscape. Clean air, and access to attractive rural landscapes were usually more readily available to regional residents than urban residents, while stallholders were keen to emphasise the aesthetic attractions of rural life – the ability to find peace and quiet, access to

national parks, and forms of recreation (such as horse-riding) not easily available
in large cities. Freedom and a sense of place were regularly enthused over; as the
President of the CWA observed:

> You're responsible for what's happening on the farm and you've got the freedom
> to make choices. I can get on the bike and go for a walk by the river if I want.
> That whole sense of freedom, looking over the paddocks and over the dam and
> seeing the cattle and the sheep, that's what I love (quoted in O'Dwyer 2008).

The notion of 'freedom' and 'escape' from the various physical and mental
'confines' of urban life was pervasive. Time was more likely to be one's own.
Further west even time and place were almost literally unconfined; Western
Queensland councils made repeated references to the all-encompassing 'beautiful
sunsets' (with not even occasional trees to obscure them). Themes of lifestyle and
scenery never disappeared but the more obviously rural, agricultural and tourism
focus of earlier years gradually shifted towards a greater emphasis on housing and
employment.

Housing

Housing, though sometimes unstated by visitors, was probably the most important
factor influencing movement and destinations, simply because it was essential.
Moreover, unlike employment, it was constantly emphasised that housing was
much cheaper and better value in the regions, relative to incomes, and in absolute
contrast to the situation in capital cities. CW never failed to home in on house
prices. As Bailey stressed, regional house prices ranged from 'under $100,000
through to a million dollars' (quoted in *Daily Telegraph*, 31 July 2009). As Sydney's
median house had just risen to over $522,000 in the first four months of 2009, this
was a valuable sales pitch. Yet, at least one Queensland council did state: 'We
have plenty of housing; it is not cheap though – we don't mention that' (quoted in
Tsioutis 2007: 35). Even so, comparable housing was usually substantially cheaper
than in metropolitan capitals (except in a few booming mine towns).

Councils gradually invested more effort in advertising housing issues and many
councils, such as Oberon, effectively teamed up with local estate agents on their
stalls. Almost every stall featured newspapers, or brochures, with house prices.
Estate agents who were able to generate interest in property or land, in places with
depressed housing markets, recognised the worth of participation. Many councils
recognised that housing issues had become more important; Yass, in 2008, could
'see a much greater urgency of people to get out of Sydney this year. Even those
with houses are anxious to sell and upgrade'. For Grenfell in 2008 'we could show
houses at $120,000 and blocks of land at $40,000 – it was our most convincing
advertisement', when the median Sydney house price was about $500,000, and

that in Western Sydney (the main source of Expo visitors) exceeded $350,000. Housing gradually grew in significance.

Employment

For most of this century unemployment levels have been very low in Australia. High employment levels partly created the need for CW, as some regions seek to fill job vacancies, while most prospective migrants sought assurances that job vacancies exist in particular destinations. As employment became of greater importance both CW and individual councils gave it greater prominence. From 2007 at the entrance to the Expo was a wall of job advertisements – the CW Jobs Board – arranged by both occupation and region, mostly in trades (Figure 4.2), while most stalls displayed their own list of jobs, often with the name of the employer, wage rate and so on. CW had a cut-off date on job advertisements of a week before the Expo, but some councils had posted last-minute additions to the list ('Cooma needs a hairdresser urgently'). These were therefore real jobs rather than notional vacancies.

In several cases visitors applied for such jobs though their value was primarily to provide an indication that jobs were available, what they were, what the wages

Figure 4.2   Employment Board, NSW Country Week 2010

were, and thus hint at some economic growth. While most councils emphasised a range of locally available jobs (as, for much of the state, did state or national organisations representing corrective services, education etc.), councils that had a specific list of particular vacancies attracted greatest interest. Smaller councils, like Glen Innes and Oberon, found it harder to advertise jobs. Encouraging the migration of couples who both wanted jobs was particularly difficult. Here too larger towns had advantages. Going from general notions of employment to specific jobs was a productive trend. One council had a direct and productive message with a placard stating 'If You Are a Plumber We Need You'.

Despite the role of real, specific and immediate jobs, generic jobs retained a place. Cardwell, with a long coastline, and a need for economic growth, was looking for people with degrees in aquaculture. Rather more frequently councils wanted anybody 'willing to work'. Innisfail thus had a self-confessed 'scattergun approach' with various possibilities for anyone interested. Every council at the Queensland Expo 2007 'clamoured for skilled workers, particularly in construction, engineering and health' (Tsioutis 2007: 32). At every Expo fitters, machinists, motor mechanics, electricians, carpenters and cement renderers were in regular demand. The Mayor of Perry Shire observed: 'I don't think the tradesmen really understand how desperate we were for them until they come to Country Week' (quoted in Tsioutis 2007: 32). While the main demand was for tradespeople, two groups were in even greater demand – professionals with such skills as medicine or dentistry, and small business owners who might employ others in turn. In Tumut 'we have to travel hours and hours to get to see a dentist. Some people just end up pulling their own teeth out. That is how badly we need them' (quoted in Tsioutis 2007: 33), which was not the message that was being offered to others who were considering migration to such towns. In 2008 a visiting dentist was strongly courted by several councils, and at least two felt triumphant that they had snared him. One of those was Tumut where 'we spent 30 serious minutes on him'. As Tenterfield pointed out:

> What do we need? Dentists! You'll find that for most country areas. We don't have a dentist! We've been looking for one for a whole year. We also need skilled and unskilled labour because there's lots of building going on here … We need electricians … carpenters … we need them all. You can wait two to three weeks to see a plumber.

What was true of Tenterfield was true of many other similarly sized towns. In Perry Shire: 'Electricians, carpenters … oh plumbers – aren't they precious?' (quoted in Tsioutis 2007: 32). Gaining new skills was a measure of Expo success. Tenterfield claimed that their entire participation would be justified if a tiler who had appeared likely to move there actually relocated. When Cooma snared a dentist the local perception was that all the money ever spent at the Expos had been justified. Filling vacancies and attracting new workers, and housing them, gradually became the core of CW.

Anti-Urbanism

Most councils formally promoted local themes and left it to discussions, or the experiences of new residents (Chapter 8), to mention the disadvantages of metropolitan life. Reasonably they realised that visitors came to the Expos out of some dissatisfaction with their existing lives. The NSW town of Junee (2009) however took a more oppositional line

> Sometimes it's the things that are missing that are the most important. In Junee you won't find traffic lights, multi storey car parks, or endless supermarket queues. You won't find toll roads, parking meters, high rise office blocks, mega malls or smog. Nor will you find Thai takeaway at midnight or round the clock pubs or doof doof coming out the door. What you will find is a small busy town just a half- hour drive from Wagga Wagga. You will find decent coffee, high speed internet connections and quality rail and road links to Sydney, Canberra and Melbourne. You will find a new hospital and medical centres, new library, great schools and a fantastic indoor aquatic centre, gymnasium and sports hall. You will find fresh bread every day and a butcher who knows your name and will soon know what you like to cook … Junee offers space for kids, and safety … In the real estate market you will find fine period homes, cottages that need love, new homes built for busy families and rural blocks big enough for those fruit trees and chooks you have dreamt about. Most of these are available at a fraction of the price of a similar metropolitan home or block.

Typically this summarised what concerned many about urban life (and its inherent pressures), offered a sense of community, continued links with distant cities and the 'best' of urban life (decent coffee). The stated absence of traffic congestion and the pleasures of short commuting journeys – and both knowing one's neighbours but simultaneously being a little more separate from them – constantly recurred.

At Julia Creek 'neighbours know and talk to each other' since it was 'without traffic or noise, litter-free, graffiti-free, no need to lock car doors and teachers can chastise kids'. Smaller populations, combined with a notion of community, spilled over into the theme of safety, security and greater freedom for children. Within CW it was almost an article of faith that crime and security were substantially worse in cities than in the country. Thus Bailey has observed that, once queries on housing and employment had been resolved:

> They will then ask about security. Security is an increasing concern for lots of people in Sydney. They want to get back to a situation – and I don't mean bombs going off in the middle of Central – I mean security in terms of being on a train or walking down the street, or letting the kids play in the neighbourhood, many things we take for granted in country communities (quoted in *Daily Telegraph*, 31 July 2009).

This too was a recurrent theme. Inland councils, familiar with urban life as much through the media as anything else, also regarded security as a problem in capital cities and therefore an important influence on potential migrants.

Only Grenfell and Narromine, relatively small and remote, somewhat unusually, had deliberately retrospective visions in featuring stability and continuity. Even Grenfell's Economic Development Officer stated: 'Nothing has changed since Grandma's day', hence their deliberations over the role of quietness, and their slogan was 'Relive the Nostalgia' (though it combined that with a tee-shirt 'London, Paris, Grenfell'). Narromine's was 'Times Change, Values Don't'. Hughenden, with 'dinosaur country' and 'not just 100 years but 100 million years of culture and history', took retrospection to extremes.

Location, Location, Location

Perhaps the single greatest challenge facing councils was the need to differentiate themselves from each other, and become the preferred destination. Location was one distinguishing factor. A small proportion of the Australian population live outside urban coastal metropolises and most Australians are uninformed, unfamiliar and unconcerned with inland Australia other than for the most fleeting of visits. It is usually out of sight and very much out of mind. Questions were regularly asked about the availability of basic services such as piped water and electricity, to the dismay of council workers, who hoped for a more informed and sophisticated engagement. When news events occur in regional Australia, maps rarely accompany them so that they happen in a somewhat generic Australia across the Great Dividing Range, around 100 kilometres inland. Grasping even a basic geography of rural Australia, in terms of location let alone economic activity, is difficult.

Colloquial Australian English language has long been replete with phrases such as 'beyond the black stump' (originally a dead tree at Blackall, 1,000 kilometres west of Brisbane) and 'back of Bourke' (a small NSW town, about 800 kilometres north-west of Sydney, whose council website describes it as a 'gateway to the real outback'). While such phrases have tended to become obsolescent in urban Australia, as regional Australia disappears from consciousness, the tyranny of distance affects how many people think, or do not think, of regional Australia. For many urban residents, a more contemporary phrase, 'West of Woop Woop', effectively defines places far away from capital cities. Councils have the initial task of persuading visitors that their towns are nowhere near Woop Woop.

Many publications highlighted concepts of distance and time. Cootamundra listed the distances from Sydney, Canberra, Melbourne, Wagga Wagga, Young, Adelaide and Brisbane. Forbes also noted their 'geographic location' by listing the distances to Sydney, Canberra, Brisbane, Adelaide, Dubbo, Melbourne and Orange. Narrabri was 'strategically positioned midway between Sydney and Brisbane, at the junction of the popular tourist routes, Kamilaroi and Newell highways'. Although some distances were enormous (with over 1,500 kilometres separating Forbes and Adelaide) they suggested some sense of proximity, to overcome perceptions that places were 'in the middle of nowhere'. Somewhat remarkably one Queensland council argued: 'You get better health care in Quilpie than you would in Brisbane. You don't have to wait in casualty for days, you are flown three hours to Toowoomba where you are treated immediately'.

Within the towns however distance was insignificant. Time was regularly invoked as a foil to the perceived disadvantages of urban traffic congestion and lost time commuting. Gunnedah had its clock display, the Mayor of Dalby pointed out: 'I can go from one side of Dalby to the other in five minutes, and it's a town of 11,000', Guyra argued that it was the '3 Minute Town', since nothing was further away than that ('and you won't have to stop at the lights or look out for the parking police – we haven't got either') and Narrabri noted 'With no traffic lights or traffic jams you are guaranteed to reach your destination stress-free and quickly, thereby allowing you and your family to enjoy a greater standard of living'. That standard of living came with adequate services – involving regular lists of schools and hospitals.

Regional councils must gain some basic recognition, where some smaller ones are unknown to most visitors and putting even the largest ones on a basic map was challenging. Only coastal towns were relatively familiar, and few of them attended Expos. That visitors have even heard of a place gave it a marketing advantage but, for the most part, visitors were confronted by the presence of stalls from about 30 different places, all advocating their own particular – or, indeed, not always particular – advantages. The outcome was uncertainty, and recognition that time was needed to make appropriate decisions, which necessitated a series of stages, sometimes represented in repeated visits to the Expos (Chapter 6). Councils recognised this: 'when relocating to the country, you are changing environments, changing jobs and homes all at once so it is a big decision that can generally be a three-to-five year plan into the future'. Continuity, the consistent presence of individual towns at Expos over a period of time, was thus essential. Visitors were challenged by the number of places, and the similarity of marketing strategies, and struggled to differentiate between them. One put it quite simply: 'I just don't know where to go to'. Others realised they actually needed to visit places: 'will there be an abattoir in the main street that they forgot to mention?' (see Chapter 6). While this necessarily made evaluation of the success of CW almost impossible, one spin-off emerged as potential places for relocation became initial tourist destinations for potential residents. At its most extreme one mother of two pre-school children confided: 'We like to visit a different place every year so that the children will know their country and we will know where to move to when they are old enough to leave home'.

Location was crucial, because more distant places were largely unknown, and relocation over shorter distances was more likely. Councils had to work hard to place themselves on real and cognitive maps. Accessibility to Sydney was touted as a virtue (at the Sydney Expos), but in Queensland larger regional cities challenged

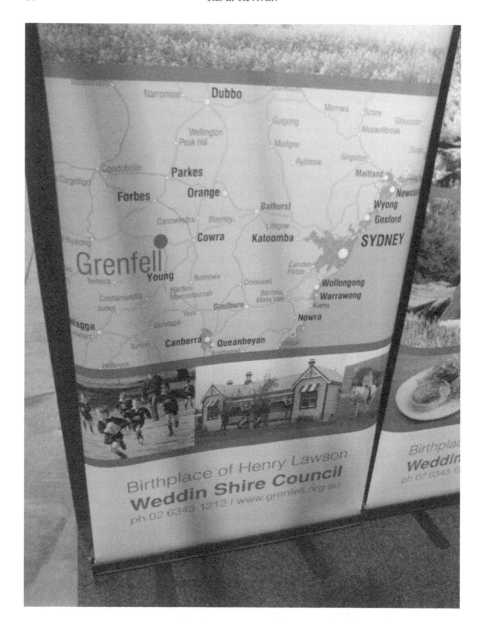

Figure 4.3　　Putting Grenfell on the map, Country Week 2008

the primacy of Brisbane. A constant refrain of 'Out of town, not out of touch' underpinned marketing endeavours and discussion on stalls. For those who feared that relocation to more distant places such as Moree would cut ties with Sydney, it was pointed out that savings on housing costs could be translated into airfares for weekends in Sydney to take in a show, sports event or relatives. The workers from even smaller Warialda, close to Moree, spent much of their time explaining where it was, though they also tried 'midway between Brisbane and Sydney'. Getting on the map was crucial, as the mayor of the Queensland town of Dalby observed: 'I've had a lot of people ask "Where's Dalby?". It's about putting Dalby on the map'. Inverell, Moree and many Queensland towns faced similar problems of both ignorance and distance and were realistic about the prospects of luring migrants from distant cities. Temora in south-western NSW had realistically concluded: 'the truth is it's very hard to sell Temora: 4,500 people in the middle of nowhere. We somehow have to get people into the south of the state'. Northern NSW councils soon recognised that their catchment area was as much southern Queensland as NSW. Thus Tenterfield attended the first Sydney Expo but never returned: 'We went to the NSW Expo and we got people really interested but as soon as they saw the map and heard about the 860 kilometres we lost them: "How's Grandma going to visit?"' Toowoomba, just across the state border in Queensland, also attended the 2006 NSW Expo, but like Tenterfield quickly recognised their real market was in Brisbane. Yet, as tiny Boorowa pointed out: 'you have to be here to be seen'; being there offered more opportunities than not being there. Many stalls had large maps, with their own town(s) prominently marked, as did Oberon (Figure 4.1) and Grenfell (Figure 4.3). Most maps had highways firmly marked, to ensure no lingering notions of bush tracks. The Glen Innes map emphasised that it lay on the crossing point of two highways – a distinct regional centre.

Regions that were relatively close to Sydney used access as a valuable strategy. In 2006, Oberon's theme was 'Relocate without losing touch', with friends, relatives and social and economic activities in Sydney. Its brochure emphasised 'We are less than two hours from Sydney (one from Parramatta) ... this means you can relocate to Oberon and still be at the centre of things'. Simultaneously it was stressed that it was easy to make repeated visits to Oberon in a 'try before you buy' process (quoted in Brown 2006: 26). For most visitors both strategies seem to have been advantageous (Chapter 6). Distant places, such as Tenterfield and Warialda, experienced problems and eventually dropped out.

Most councils sought to convey distinctiveness alongside generic positive images (employment opportunities, low housing costs and community), but only a few regional centres were even tolerably well-known. Tamworth is known to many for a longstanding annual country music festival, though that image is not necessarily welcome to all in Tamworth (Gibson and Connell 2011); some council staff saw this as a potential liability in competition with nearby places such as Armidale where a university might attract more educated migrants. Parkes too had a distinctive Elvis Presley Festival, but most festivals reached only local communities. Few places had otherwise acquired any state-wide recognition. Consequently many visitors inquired about larger towns such as Bathurst and Orange, that are close to Sydney and never came to the Expo, or coastal centres. Other towns had either sought to create distinctiveness, such as Glen Innes or Moree, or confessed an inability to achieve it; as Muswellbrook confided: 'We are not sure how we can adequately distinguish our town. We emphasise jobs and water but who doesn't do that?' Perhaps just as confusing for visitors was that the stall workers were entirely convinced that their own region was superior, though competition between regions was generally good humoured.

Taking the City to the Country

A widespread assumption existed that visitors, potentially making a transition from large cities, would need convincing that some metropolitan facilities were available in smaller towns. By 2009 CW were even claiming that 'where the Big Mac index has become an international measure of affordability, regional NSW can now claim a Cappuccino Index for liveability' (*Daily Telegraph*, 31 July 2009). State Government Departments were not averse to similar notions, though their promotional material may have been written in Sydney; thus the CW edition of *Hunter New England Health Matters* began an article on 'Going through a tree change' with 'The bakehouse in Murrurundi supplies ciabatta to Neil Perry's Rockpool and the wool spun in Nundle is crafted into bold Ken Done winter apparel for duty free retail at the Rocks'. (Rockpool is one of Sydney's most exclusive restaurants and the Rocks is the main destination for upmarket tourist shopping). While Junee offered good coffee, but rejected Thai food at midnight, the notion that having one's cake (and cappuccino) and eating it – the best that rural and urban can offer – was entirely possible in regional Australia. Though it was rarely a significant theme in council promotions, advertisements for good restaurants and high quality stores were prominent in brochures from the larger towns. But such 'new cosmopolitanism' was, literally, not to everyone's taste. As one established resident in the Victorian town of Castlemaine complained: 'There are now too many places you can buy coffee in Castlemaine. It's become more visitor focused' Another declared that it 'could be just a suburb of Melbourne as far as the café latte set and everything else that goes with it' (quoted in Costello 2007: 92). Change could always be construed as being disruptive and marketing needed to account for a range of perspectives from both local residents and potential visitors and so err on the side of the bland and most obvious.

What Lies Beneath

Necessarily councils emphasised the positive. The special CW editions of newspapers stressed the multiple charms of particular communities and the success stories of those who had moved there. Regional life is not however without its problems, some of which are inherent in distance and size, which necessarily reduces access to certain services, and may be obvious to thoughtful, potential residents. However, other 'problematic' features are neither obvious nor discussed. At one level this involves physical infrastructure – both positive (various services) and negative (the dreaded abattoir in the main street) – alongside concerns that the climate might be too cold or too hot. Tumut, high up on the western flanks of the Snowy Mountains, had their own climate table demonstrating to sceptical visitors that it very rarely snowed there ('only once in ten years' – though it snowed on the first night of the 2008 Expo) and that it was not far enough south to be too cold. Nearby Cooma's 2010 tee-shirt proclaimed 'Cooma: cool not cold' (and their leaflets offered a 'ski-change'). In Queensland climatic issues were somewhat different; Bundaberg positioned itself as 'south of sweaty, north of cool'.

Many inland towns have significant indigenous populations, and in most towns Aborigines are more visible than in Sydney or Brisbane. In both NSW and Queensland respectively, the majority of indigenous people live in regional locations, in or near many of the towns that attend CW. In some of these towns relations between Aborigines and others have not always been easy and relatively high levels of unemployment and low life expectancy and high morbidity characterise indigenous populations. Several towns have experienced occasional racially related violence and crime, partly linked to racial discrimination. Poor race relations have given a social stigma to some towns, such as Dubbo, Moree, Wilcannia and Bourke, and may deter potential new residents if dwelt upon. Most visitors from metropolitan Australia would have had limited knowledge and experience of local indigenous issues. Few stalls hinted at an Aboriginal presence in their towns though Lachlan Council had a brochure promoting the Wiradjuri Corporation (a public company promoting 'local Aboriginal self determination, economic and social independence'), and Liverpool Plains and Cootamundra acknowledged their shires as the 'Home of the Kamilaroi people' and 'part of Wiradjuri Country' respectively. Even the tourism literature largely excluded Aborigines. In NSW over a four year period no stalls appeared to have Aboriginal staff. The lone and quite distinctive exception was the Torres Strait stand in Queensland, for a council where the population is almost entirely Melanesian Torres Strait Islanders. The stall was staffed by both Islanders and Europeans, and at regular intervals featured distinctive Torres Strait dancing displays. The primary objective of the Council was to stimulate tourism rather than migration.

Branded Diversity

Evocations of variants of countrymindedness initiated most place branding but over time the orientations and strategies of councils evolved from a more touristic focus on the aesthetic to a more prosaic yet vital focus on employment, housing and other services. In regional Australia they coincided almost exactly with those used to promote the attractions of small declining industrial towns in regional Sweden: 'good communications, low costs for houses, good quality of schools and services and an attractive natural and cultural environment' (Cassel 2008: 111). Such broad similarities indicate that even flexible place branding must be basic, straightforward and 'consistent and cannot allow for polysemy, plurality or contradiction without the risk of becoming an indecipherable cacophony' (Mayes 2008: 127). Evolution of branding strategies proceeded differently according to finance, motivation, analysis and comprehension. Councils that had more information available were in the best position; councils, like Mt Isa, that did 'not want to bombard them with too much information' were more likely to be ignored. Bombardment sometimes worked, but even good information could easily be cancelled out by remote location.

Some councils decided to attend only a couple of weeks in advance; others had mere fragments of information available. Still others had spent months in preparation. In 2008 the small town of Crookwell, somewhat disguised under the banner of Lachlan Shire (a title that would mean nothing to most Sydney residents), had no available information on local employment (though the Council itself was short of at least three skilled workers). Effectively those working on the stand were reduced to making vague statements about access to water, jobs being available and so on, without the precision that had evolved elsewhere. Lessons were slowly learned: in 2009 a stall worker who had moved between council areas commented that 'it was much easier to "sell" Grenfell since there we had actual jobs; here it's hard to pinpoint actual ones, though we are probably short of everything, so we can only make promises'.

Most councils recognised the need to be prepared for even the most asinine queries, especially where towns were remote or unfamiliar. Normality had to be demonstrated; 'City people think we are hicks. Hicks from the sticks we are not', 'We don't want to appear as "bushies" though. We are just normal. We don't have four eyes, we're just people'. Moree found that in the first year they were being asked such questions as 'Do you have electricity out there?' and 'Are there flushing toilets?'. In response their presentations and brochures became more detailed and more upmarket, yet similar questions about water and electricity were asked in every year. At least as difficult was the task of marketing their places from anonymity to attraction, in the face of state-wide competition. As such questions indicate the images and realities that councils have sought to promote are not necessarily synonymous with the images and aspirations of visitors to the Expos (Chapter 6). Nonetheless the CW Expos were one of very few opportunities for inland towns to demonstrate that they had valuable qualities, and to move people away from perceptions that 'rural' was negative. Indeed here there was commonality alongside competition. As one council observed: 'it doesn't matter if they don't come to Richmond as long as they get a sense that there's good living and lots of opportunities out in the country'.

Participation at CW Expo required a significant commitment of resources, plus important preparation and follow-up, discussed in the following chapter, in order to be successful, to boost local identity, encourage migration or generate tourism. The scale of the event demanded both cooperation (between the private sector and the public sector, in particular places) and competition to construct a distinct brand of 'country' that might embrace diversity and ward off competition from elsewhere.

Chapter 5
A Place on the Map?

We have to be careful we don't come over as used car salesmen (Council Worker, NSW Expo 2007).

Six years ago no-one had heard of us, three years ago they knew where we were, now they are coming (Oberon council worker, NSW Expo 2010).

One of the most difficult things for Country Week (CW), and for participating councils, is measuring the success of the Expos and, beyond that, evaluating why people move to regional areas, and how such moves might be used to develop superior future strategies. It is however a characteristic of place branding and marketing strategies that very little is ever known about the actual outcomes (Niedomysl 2007). This chapter examines the extent to which it is possible to measure the outcomes of the Expos. While the number of visitors to the Expos is an initial measure of success, these numbers include 'tyre-kickers' – prone to collecting showbags and wasting time – and only some visitors even consider regional mobility further let alone move. For the councils the Expos themselves are part of a longer process that includes evaluation and engagement with contacts and other follow-up activities to assess what may or may not have worked.

Building Community

Despite uncertain outcomes, many councils believed that attending CW was not just a possible means of acquiring new migrants but brought people together to consider what issues their towns should focus on, as they developed marketing strategies. Through bringing together such people as economic development and tourism officers from the public sector, estate agents from the private sector, newspaper publishers, mayors and councillors, a sense of community could be developed and revitalised: 'internal marketing and identity formation'. Such ancillary benefits were invaluable. CW was even seen as a means of removing people from their communication 'silos' and engaging in a shared, positive activity that would enhance teamwork and yield various beneficial results. Promoting communities elsewhere potentially improved internal cohesion. For one council: 'If you are going to solve some of the tough inward problems it is very important to get people to focus outwards. They start to think of ideas, to get recharged, and this feeds back into the community'. The Expos thus promoted towns to their existing residents, enhancing identity through evoking a sense of pride, action and achievement.

Some councils worked extensively with local employers and media prior to, and following, CW. This enabled them to have more specific and contemporary information available at the event:

> In the lead-up the Mayor talks quite often to the media, we put out media releases about going to Country Week, we have a good relationship with the local newspaper and they're very supportive of the exercise but unless you are actually advising and bringing the community with you to show that this is what we're doing, we're not just doing it for the sake of the Council having a weekend away -this is what we are trying to do to attract people and services to our shire and that will ultimately benefit you.

By 2005 Oberon had set up a team including council members and representatives of business, industry and the tourist organisation to develop a 'skills audit', involving questionnaires to local residents and businesses, so that local businesses could advertise for and attract the most appropriate skilled people to Oberon, enabling it 'to grow and prosper as an independent community' (Oberon Council 2005).

Many councils stated that the decision to attend was conditional on how well that particular Expo went. For example: 'the crowd is down ... If we get a few nibbles from this year I think we'll be tempted to come back, but if we don't get anything from this year I think we'll be reassessing our commitment'. Another council noted:

> Once we've done Country Week we then go through a process of debriefing about what we saw and what we need to do. [Council officials photograph] all the displays to evaluate what they see as being beneficial and what is not. So it is going through that continuous critical analysis right to the point where ... about six months before the next Country Week we will go through that process of what we need to do.

Some councils saw CW as part of a bigger commitment to marketing their town, but for others the Expo was the main marketing tool. Other events such as follow-up tours (see below) were closely linked to it, with subsequent activities being impossible without the CW 'gateway'.

> I think the idea is great and coming down here is a better idea than holding an open weekend when we try and say everyone come up here, because we don't have the contacts. We don't know who the people are that are interested. We need to contact them in this environment and then invite them up.

Estate agents increasingly found the Expo to be particularly important for them, and so provided financially support to council participation. Sometimes importance was related to timing. A particular skills shortage, an impending boom, or a threat to the continuation of services, gave CW greater importance.

Being There

The actual experience of being in the city and working at CW Expo was generally viewed positively by councils, especially as they became more aware of what people were looking for, and thus were able to plan branding and promotion activities to maximise the benefits:

> First year was a 'suck it and see'. We had no idea of the reaction, no idea of the number of people who were going to turn up, all those variables … It gets easier as the years go on as far as knowing what to expect. But, there's always preparation in developing up a theme, knowing what promotional material you need to develop because of the theme.

One attraction of the Expo was that councils were making contact with people who had deliberately chosen to attend this event, and not usually wandered in merely because they were bored or passing by. Councils listened to their enquiries and consequently refined their messages and marketing strategies:

> We took the feedback from the past. Jobs and employment is first and foremost. Then it gets down to housing, education, health, lifestyle and location. They're the six items.

> We've matured our marketing that we now focus on three or four major themes – housing, the availability of homes and rentals or whatever, job prospects and the overall lifestyle issues like health, education, coffee shops and leisure as well. [In the past] we were working on quirky themes … we soon realised with the questions people were asking how we needed to focus.

> There are four main factors that people ask for in their decision making process. The first thing is, will I have a job? If they've got kids, what sort of home can I afford? Children, then there's education and health, those essentials of life. Can they be provided to an acceptable standard? Finally, lifestyle, that's probably the easiest one we can sell. In our town you have more time to spend on yourself and your family because you are not sitting in peak hour traffic.

Any decision on attendance was influenced by the available resources, and how the council wanted to project itself. Particularly important was local media support and, through them, community support for the council's expenditure of rates. Several councils had changed the composition of their delegation over the years for better communication with visitors and to involve a diversity of local interests:

> In terms of what we did this year we didn't change a great deal. What we did
> change was some of the personnel involved. We continue to widen it in terms
> of community involvement, so you've got the business interest and real estate
> agents who actually come in, we've got representatives from the teaching
> profession. It's moved away from Council being the dominant participant to
> being the facilitator and organiser ... There's more to our community than the
> council.

Changes in personnel sometimes generated changes in presentation. 'This
year we have gone a lot more casual ... because this year we have community
representatives with us as well. We didn't think it was fair to ask them to go with
our corporate image'.

Media coverage was both a concern and an opportunity for CW and the
councils. The national media were given releases about topical issues during the
event, but concern over the city-based national media was widespread:

> We know all the stories that come out of the city media – as soon as there is
> anything that says doom and gloom they immediately latch onto it. In the 1990s
> they talked about bank closures, they talk about drought, floods, all the natural
> disasters, and the fact that country towns are dying. You hear it all the time. We
> think then that the perception of city people is, why would you move, because
> there are all these terrible things happening out there.

Councils obviously challenged this, at any possible opportunity, while their strategic
evolution resulted in a growing focus on issues that had been 'difficult' previously,
as they sought to directly confront negative perceptions. However inaccurate were
'perceptions of the dry, dusty country areas of inland Australia, [they] hamper
our ability to attract new residents' so could not be ignored. Similarly councils
became increasingly aware of the need to target particular groups who were more
likely to move, in part because of their greater familiarity both with country and
with some particular places. As one CW staff member observed: 'That's part of
our message. The research we have seen before is that people are more likely
to go to the country if they have had some sort of country experience. That is,
either they came from there, or were educated here'. Those without links to the
country were less likely to become residents. Recognising that neither words nor
brochures were in themselves convincing, and that some familiarity was crucial,
councils increasingly emphasised the importance of visiting the town or area:
'We're stressing to people to come up, have a holiday, have a look at the area
and hopefully come back and stay'. Visiting necessitated no commitment and was
important in supporting rural hospitality businesses.

While councils quoted with both pride, and some amazement, rare examples
of the occasional Expo visitor who had moved to their town (or bought a block of
land) solely as a result of visiting the Expo, such moments were exceptional, and
migration was an extended process. Councils however often claimed that a number

of households had migrated as a result of visiting the Expo. Armidale claimed at the end of 2007 that 11 households had already moved as a result of visiting that year's Expo. Oberon claimed 'two or three households' moving there each year as an outcome, but the Expo can only be one influence: a possible catalyst but not normally the sole cause of migration.

Councils recognised that migration would take time and due consideration so that their own repeated attendance at CW was important, as visitors returned in subsequent years, perhaps narrowing down or finalising choices. Moreover migrants from Sydney, at least, would move on the assumption that they could never get back into the Sydney housing market again (a realistic assumption since metropolitan house prices have consistently been higher and risen faster than those in regional areas), and visitors often move at significant points in their lives (such as when their children leave school). Oberon council backed this up by pointing out that one household had visited their stall on the two previous years and had returned triumphantly the third time to tell them that they were about to move to Oberon (Brown 2006: 28). Their 2010 stall was staffed by seven recent migrants, a tradition for Oberon, some of whom had been encouraged by previous Expos. Similar perspectives inform the strategies and presence of other councils.

Open Days, Tourism and Festivals

It is axiomatic that visitors to CW will not move to regional centres unless they have actually seen them. Getting possible migrants to visit and appreciate the towns was therefore crucial. Once again the primacy of location was evident. As Moree, one of the more distant local councils, noted: 'Our biggest problem is to get people here since we are so far from Sydney. Once they are here we can show them the community'. In 2007 the Moree mayor was giving people his card and telling them to contact him when they came to Moree. A year later the Council developed the idea of the 'Healing Waters Express', an art deco heritage train that would bring Sydney residents to Moree to experience the spa waters. Four free tickets for this were given away as prizes at the Expo stall. The train duly ran in October with 25 passengers, less than anticipated, and incurred a loss of over $9,000, although initial reports put the figure much higher (Moree Plains Shire Council 2009). Such outcomes might have explained Moree Plains' absence in 2009.

A handful of towns had subsequent 'open days' specifically aimed at potential migrants using a weekend visit partly to assess residential potential. Visitors to Grenfell's Open days were offered a $50 petrol voucher and a $25 gift voucher, and 12 months' free rates if they relocated, with 'additional incentives for skilled tradespeople'. Cootamundra and Muswellbrook similarly gave out petrol vouchers. Dalby provided free lunches, barbeques and bus tours of the town. Grenfell claimed in 2007, with perhaps dubious mathematics, that 'about 50 percent of the people who visited Grenfell on the Open Day ended up moving here. We have had about five families move here since we first attended … We've had a plumber, a carpenter … an electrician … it's been great'. Perhaps uniquely Oberon took its show on the road, along with estate agents, and held information nights in different venues in western Sydney.

All stalls had some standard local tourism literature and information on festivals; some had specific promotions with holidays as prizes. In most councils the Economic Development division was usually linked to Tourism (and workers from these two divisions were the main official staff at Expos). Tourism and festivals are partly synonymous, and few if any towns are without them. Festivals were an important part of marketing, partly because towns are seen at their best at such a time – using a particular season and the weather as attractions and hinting at various forms of cultural life and activity. Inverell, for example, displayed fliers for its annual Opera in the Paddock – which draws over 1,000 visitors to Inverell for the one-day concert – a clear indication that 'high culture' exists in the countryside. Conversely Parkes has a tourist office with several staff, one of whom is specifically engaged in developing and promoting the annual Elvis Festival. Like the Elvis Festival, the Glen Innes Celtic Festival draws about 6,000 visitors to a town with about the same number of residents, minimally representing a substantial boost to the local economy, while the annual week-long Tamworth Country Music Festival is the third most important event in Australia in terms of the numbers participating and the income generated (Gibson and Connell 2011). Such festivals draw visitors, some from metropolitan areas, create employment and effectively promote regional Australia. Some such short-term visitors may return, even permanently. Through these festivals, potential residents become familiar with towns, and regional areas, and are better informed when making decisions about relocation.

Following Up

The need to directly follow up the most promising enquiries was seen as crucial by regular council participants, but it was not always possible because of staff changes or resource constraints. Without effective follow-up, part of the time, effort and money put into being at the Expo may be squandered. Data management was crucial:

> This year we are concentrating on achieving more precise data catchment while we are here. More importantly, more businesses are involved in preferential offers for job-seekers who want to visit [our town]. We have accommodation packages, meals at restaurants, either at a preferred rate or having some sort of benefit, to translate the registration of interest here into actual visits.

> It is important to keep up contact with them, to provide them with further information, to invite them to [our shire] and once they are there the process we follow is, although there's lots of councillors and people who may want to go on

the bus tour, we make sure that the people who have been at Rosehill and have been fielding the questions and understand what these people want to know are placed in a position of providing them with the information.

The most direct follow up was writing to those who had shown some interest and had been registered by councils. Typically this went along the lines of the 2007 letter from Pine Rivers Shire Council:

Pine Rivers – Where Lifestyle Meets Opportunity.

Congratulations on choosing to visit the Pine Rivers Shire Council stand at the Queensland Expo. We enjoyed meeting you there and having the opportunity to tell you why we think Pine Rivers is a great place to live. We would like to try to help you make the move to Pine Rivers and we will be happy to assist you where we can. There are many opportunities in the Pine Rivers Shire, as we mentioned while we were talking. We are prepared to forward your details to businesses and organisations looking for your specific skills. If you would like us to proceed with this, please forward your current résumé so we can give your most recent details to employers and agencies looking for your skills. If there is any other information about the Shire you would like, forward your request and we will find it for you. We hope to see you here soon.

Parkes similarly sent a letter to about 30 people each year, those that they regarded as the most serious visitors, and in turn half of those sought more information (though the final outcome, as usual, was unknown). Councils often had an ordering of priorities for follow-up, which either matched the most promising enquiries, or reflected the town's most important needs.

You want to follow-up with jobs and job seekers first and foremost. If there is significant inquiry on retirement ... if individuals are going to be moving there anyway through their own preference we've got to listen to that. Likewise if there is tourist inquiry.

The main thing that I will do different is rank the inquiries, and follow-up those ones ... that we know are suitable. With those people we will follow up with phone calls, emails and information packs. With some of the working ones what I'll do is time the email mid-afternoon so that when they've possibly had enough of work ... I'll send out a new email each month. The process is that these people are not going to move tomorrow.

Councils thus recognised that while moving was a long-term process, the timeframe for effective follow-up was quite limited.

The phone was ringing with the odd phone call from people who had registered
an interest in the two or three months after Country Week. I can't recall a single
phone call this side of Christmas [four months later].

Councils consequently tried to attract people to visit the town in the months after
CW, in some cases before it became too hot.

We'll be following up with another event ourselves at the end of October –
Country Houses Expo, and that will have real estate agents, progress associations,
schools, local builders, that sort of thing and what we are going to do after we
scan people in today is that I will be contacting them after this event, send them
a letter, invite them to a barbecue, meet the councillors, meet the mayor, have
a look at this Expo we've got and I also intend to have open houses with real
estate agents on that day.

CW was not the only means of encouraging migration to regional Australia.
Most councils had economic development officers but their tasks, as in the case
of Oberon, were 'identifying growth industries, developing a tourism strategy,
promoting facilities to community groups and other users and to work with the
business community to establish strategies to foster job creation in the area'
(quoted in Brown 2006: 29). They were thus focused on business and industry.
Moreover there is no other forum like the Expos to bring economic development
officers, and their development strategies, together with people who might help to
implement them. But the Expos were primarily oriented to households with vague
and more diffuse goals.

Evanescent Outcomes – Defining and Measuring Success

Evaluating the success of CW is almost impossible. A brief visit to a show is
unlikely to instantly turn an urban resident into a definite migrant, and CW staff
sought to communicate this to participating councils. Most visitors had never
seen or even heard of many of the places that were present. Securing a place on
mental maps was a crucial first step. Ensuring that visitors take at least another
step forward was thus a key theme, but the Expo was only one of many stages,
influences and turning points on the way. Business migration, where employment
creation and income-generating effects are more visible, was particularly difficult
to achieve. Even small companies feared that markets might be different and
too small in regional areas (where their market is not a national or international
one), appropriate workforces would not be available in country towns and/or
that their existing workforces would not want to migrate with them. Some feared
resentment from established businesses in areas where populations were scarcely
growing. While households do migrate to the country, the role of CW, the Expos

and council follow-up activities is impossible to determine, though the limited evidence suggests that it is slight (Chapter 7).

Even the impact of the migration of a single household to the country cannot be measured since households vary enormously, hence the costs of attending the Expos are hard to measure against the benefits. That did not stop councils making optimistic projections. Warwick's Economic Development Oficer suggested: 'In the next 12 months if we've got 20 families, and 20 families in 20 houses ... which is $10 million in housing stock – 20 families is 80 people, that's $60,000 per head'. Oberon's consultant at CW argued that it only took 'a couple of people moving each year to Oberon' to make attendance worthwhile since if that couple earned about $100,000 a year then a significant proportion of that amount was likely to end up in the local economy and benefit the town (quoted in Brown 2006: 27). However that was a rather generous earnings estimate, with retirees' incomes and expenditure being much less, while 'leakage' to larger centres such as Bathurst or Sydney is almost incalculable.

Councils sought to find successful outcomes where they could, not least by stressing successful migration in their promotions (Chapter 8), but well recognised the difficulty of defining and measuring success

> We got together and seriously considered whether we would participate this year... It's complicated by the fact that it's very hard to measure success. We're talking about people making decisions maybe in five years time they would like to move. How do you do the cause and effect of who came to Country Week, who moved to our town?

There is no requirement for people to register their move, let alone why they have moved, hence councils often relied on the advice of estate agents about people's relocation decisions.

> We'd like our real estate people to let us know if they sell a house, in say the next six to 12 months, we'd like to know about it so we can contact those people and give them a new residents' pack and maybe a little survey asking them how did they find out about our shire, why did you move?

Oberon Council surveyed local estate agents quarterly to assess mobility. Some instances of achievements could be reported:

> Last year we saw two young couples ... they were desperate to get out of the Navy and get a job and both of them were in a trade area and both of them came up to the council chambers and introduced themselves and said that we were at Country Week last week and now we are here, and we have got a job and we are going to live here.

More frequently success was speculative and based on optimism:

> The first year we had two families move to our town because of Country Week. Last year ... I'm not aware of anybody who has specifically moved at this stage, but we did have a couple of really positive leads.

Most councils reached similar intangible conclusions. Much like Parkes and Grenfell: 'if we get five or six real people that's a boost' and 'if we can get three households out of it we've done well ... losing a class and therefore a class teacher is a disaster', but there was no immediate way of knowing. Success was also measured in supporting the development of alternative industries to diversify the local economy, though that was not the objective of CW or likely to be an outcome of it. A few councils discussed the importance of developing industries that could provide an alternative economic base beyond agriculture.

> There will not be a repeat of the major investment in timber manufacturing ... that means tourism will probably be back as one of the major industries of significance in three to five years time. There will be a far greater emphasis on tourism.

> Our town is based on agriculture ... over 150 years the town's economy has fluctuated according to the whims of the seasons, the markets and government policy. One advantage that coal, or any other significant industry for that matter, brings to an area is that it can put a base in a town's economy.

In keeping with the shift to a post-productivist countryside, councils invariably saw tourism as one crucial opportunity, sometimes linked to the significance of festivals. Even councils such as Gunnedah, where mining was the most important economic activity, had appointed tourism officers and developed new brochures. A rise in tourist numbers as one outcome of the Expos was eagerly anticipated, but there was no real evidence for it. There were promises: 'The Expo has inspired us to travel extensively in NSW'.

Ultimately measurement of success was closely related to expectations. Councils that had limited, but perhaps realistic, expectations about outcomes were likely to proclaim themselves successful: 'Success for us is if we could get a couple of families to move. A business to move would be a huge success'. Perhaps not surprisingly, the same council noted that attendance at CW was 'a year to year proposition' and that it had to be justified, but justification was no easy task. One measure of success was perhaps whether councils could somehow justify funding for another year. Councillors and mayors often attended and could see for themselves the size of the crowd, and the level and kind of interest in their particular town. In the end it often boiled down to the simple idea that 'It's being at the market-place. It's a cost-effective way of marketing our community to an audience that is motivated to be here'. While costs were tangible effectiveness was much less so.

A Part of a Process

CW Expo is a major and costly event for many local councils in NSW and Queensland. Financial costs, plus the time and energy involved in preparation, participation, follow-up and evaluation, and their opportunity costs, raised questions about the benefits of promotion, whether there were more effective mechanisms and whether existing promotion worked. Many councils returned annually to the Expos, simply because it offered something unique: a forum for councils, state government departments, private businesses and other organisations to gather in one place and promote the idea of moving from the city to the country. This was what visitors expected, and were attracted by being able to visit the stalls of numerous places in a short period of time, a few hours 'out in the country'. Alone councils could not reach such people. At least some visitors were in 'the right frame of mind' and not there to 'kick tyres'. Many councils had thus gone from feeling they needed to 'be there' and to 'be it to win it' to recognising that effective policies and promotion took time, but that developing them strengthened community and identity.

Councils that were most successful, relative to their inherent advantages and disadvantages, were those that worked well with the local community beforehand, had thought carefully about their personnel and branding, engaged in effective follow-up, monitored their performance and participated over a period of time. Councils with expectations of instant and quantifiable success were disappointed. CW is part of a process. People do not move immediately to regional Australia, hence the continued presence of councils at the Expos was vital for the event itself, provided internal benefits for regional towns, and was appealing to visitors who sensed that a council and its representatives were solid and genuine, and therefore to be trusted (Chapter 6). While councils recognised that instant successes rarely followed immediately, patience was at a premium, attendance was costly and councils had to be responsive to voters and tax payers. To speed up the process councils encouraged visits by potential residents, for open weekends, holidays or festivals, to be proactive rather than passive. Rare councils have gone further. Oberon had contemplated packages where the Council might rent cheap houses to skilled workers for six months (Brown 2006), while rentafarmhouse (Chapter 3) was a variant on this.

CW alone can play only a tiny part in 'rebalancing' Australia. A brief attendance at an annual weekend event will not significantly influence counter-urbanisation, but the Expos provided basic information and played a small part in transforming perceptions of regional Australia. Councils were happy to tell stories about instant successes, like that told one year on the Grenfell stall:

> This man was at the stall looking somewhat puzzled. 'I can tell you're wondering where Grenfell is' I said. 'Yes' he agreed, so I explained where it was and what it was like and within a month the man and his wife were living in Grenfell and we were delighted to have another electrician.

Such circumstances were exceptional, and possibly apocryphal. Far more often residential change was a slow process that took repeated visits (to towns and Expos), and major family debates and discussions, until the eventual catalytic moment. It was not the outcome of an hour or so at an Expo. However over time councils became more adept at recognising those for whom relocation was more probable and with whom discussions and follow-up were worthwhile. Who actually attended the Expos, why they were there, what they were anticipating and how they perceived the country can now be examined.

Chapter 6
Going to the Show

'You can walk rural New South Wales in a day.'

Peter Bailey, 2006

The success of the Country Week (CW) Expos largely depends on who goes to the shows, what they gain from them and how they respond. Despite claims by the organisers and some councils, many visitors to the Expos have no definite intentions of moving, and often little more than a vague sense that the country might have something to offer and that a pleasant afternoon might be spent browsing around the stalls and finding out something about different places. A handful of people were there merely to collect whatever was free. A rather larger group were there for a second or third time, as they sought to build on early impressions and firm up what they had already thought and learned. Each Expo has attracted thousands of visitors; although the exact number is hard to quantify, CW estimate that they have had 3,500 to 4,000 visitors in most years. This chapter examines who these visitors were and what they gained from visiting the Expos.

While the number of visitors has scarcely grown over the years, many individual stall holders were anxious to emphasise 'quality rather than quantity', and that visitors had become more committed to moving, and more evidently asking realistic questions, that hinted at mobility rather than accumulating showbags. Moreover, according to the councils, there were fewer 'tyrekickers' and more serious participants at CW Expos than at other promotional events, such as retirement and lifestyle expos. While the CW Expos provided some entertainment, they were not focused on rides, competitions, farm animals or a carnival atmosphere: exhibitors saw a serious focus as desirable. For many visitors the Expo constituted something of a voyage of discovery, hearing about new places, learning about places they knew little about and creating a new geography of their state.

The Visitors

The most detailed data on visitors came from the 2007 NSW and Queensland Expos, supplemented by surveys at the 2006 and 2008 NSW Expos, and observations at the five others. Broadly similar patterns were evident at each, despite the Queensland Expos being more focused on employment. Most visitors to the NSW Expos came from the nearby western suburbs of Sydney where advertising had been directed (Chapter 4), and in Queensland they also tended to come from Brisbane suburbs of similar, more blue-collar socio-economic composition, but biased towards the city centre where the Expo was held. They came from a range of occupations, from a small number of people in managerial positions to nurses and teachers, tradespeople and the self-employed, alongside retirees. Few were unemployed.

At both the 2007 Brisbane and Sydney Expos the largest groups were over-50s (over a third of all visitors at both Expos), while under-25s were poorly represented (less than 10 percent at both Expos). Some two thirds of all visitors were in employable age groups. The largest groups at both Expos fell into the 'mother and father and children living at home' category with more than a third of all visitors at both Expos, while couples without children were almost as frequent: 26 percent in Brisbane and 35 percent in Sydney. By contrast there were very few single parents (4 percent and 3 percent respectively) and few people living alone (9 percent and 10 percent respectively). The Expos thus attracted typically nuclear family groups, but including older couples whose children had moved on or younger couples moving towards parenthood.

Predictably the majority of visitors (56 percent in Brisbane and 63 percent in Sydney) were 'considering moving to the country', a minority were 'interested in learning more about the country' (29 percent in Brisbane and 22 percent in Sydney) and the remainder – 15 percent in both cities – had already 'decided to move to the country'. At Sydney in 2006 some 27 percent had already 'decided to move'. While such crude categories can be variously interpreted, most visitors to the Expos were very far from being definite migrants, but remained to be convinced and primarily wanted more information. Two groups of people were however particularly interested in moving: firstly, younger, often recently formed households, who were particularly concerned about the cost of housing and considered that country towns might offer a better lifestyle for bringing up children, and, secondly, older people whose children were now independent, whose parents did not require significant amounts of care, and who felt able to realise ambitions to move out of the city in search of a better lifestyle. The first group emphasised economic issues; the second were less constrained.

The expressed rationale for a potential move to the country varied significantly by age, most evident in 2006 where housing and employment were of major significance for younger groups but of reduced significance for older visitors (Figure 6.1). For people over 50 urban housing costs were no longer of significance but more general lifestyle gains had become more significant. There was neither overwhelming dislike of big cities nor any pervasive fear of crime. In 2007 a broadly similar picture emerged from both Expos, though in much larger Sydney (4.1 million in 2006 compared with 1.8 million in Brisbane) distaste for urban life, and urban house prices, was greater. In both cities the positive factors that drew them to the country included a friendly community (73 percent of respondents in Brisbane and 66 percent in Sydney), the 'attraction of the outdoors' (68 percent and 63 percent respectively), the idea of a new lifestyle (over 60 percent in both) and good housing (more than 50 percent in both). Where many were casual visitors rather than determined migrants this focus on lifestyle was a likely outcome. Nonetheless, from a sample of 466 people, the cost of housing was rated by over 88 percent of visitors as of at least some importance, while access to

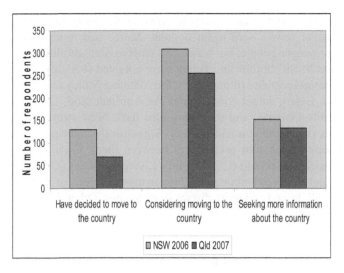

Figure 6.1   Expressed influences on migration, Country Week 2006

employment was similarly important for 80 percent of the visitors. By contrast generalised concerns over crime and urban life were of limited consequence, though asking specific questions about particular factors tended to reinforce their importance (Tsioutis 2007: 60-63). Again there were significant differences between age groups. While people who wanted to be in the workforce for a number of years emphasised employment opportunities, it was of little relevance to retirees or people contemplating imminent retirement. In Brisbane alone a small but significant number of younger people (aged less than 25) were interested in moving to the country to make a 'fast buck', earn 'big money' and the move back to the city for the urban lifestyle: an indication of the significance of the mining boom in Queensland.

These ratings of factors were undertaken by Expo visitors, most of whom were only beginning to consider the possibility of movement, many of whom had just visited their first Expo, and perhaps saw the country though somewhat rose-tinted spectacles. Nonetheless their responses indicated a very strong preference for a kind of outdoor, community-based lifestyle that appeared to have receded from metropolitan life, though few were discouraged by supposed urban anomie, but simply felt there were better options elsewhere.

Particular towns and regions in NSW were favoured for relocation (Table 6.1), and this too varied somewhat between age groups. Visitors were influenced by the presence or absence of particular councils, and by the displays, messages and general reception given by their representatives. Younger visitors in the 25-35 year old category were interested in a variety of locations, but preferred larger towns with diverse services (e.g. Armidale and Tamworth), and places that provided definite employment opportunities (e.g. Tumut, Oberon and Muswellbrook).

People over 50 years tended to want to move north, partly for climatic reasons. The same larger towns were again popular. Moree had more appeal for older people than younger people, but a reduced emphasis on employment meant that towns such as Muswellbrook and Tumut were not rated so highly. Overall visitors preferred coastal locations (though only the Manning Valley and Port Macquarie were at that Expo), larger towns (such as Armidale and Tamworth), nearby places (notably Oberon), and they may also have been swayed by impressive presentations (Cootamundra and Moree). Significantly eight places that were at the Expo secured just four preferences between them: Glen Innes, Gilgandra, Gwydir (Warialda), Lachlan (Condobolin), Liverpool Plains (Quirindi), Narrabri, Narromine and the Warrumbungles (Coonabarabran); each of these places were relatively small, inland and quite distant from Sydney. Other than Glen Innes few appeared regularly at the Expos thereafter.

Table 6.1   Preferred places for relocation, NSW, 2006

Location	Number of Respondents	Location	Number of Respondents
Albury	9	Moree	9
Armidale	25	Mudgee	8
Ballina	*3*	Muswellbrook	7
Bathurst	2	*Newcastle*	*2*
Blue Mountains	2	*Northern Rivers*	*2*
Boorowa	3	Oberon	14
Canberra	8	Orange	5
Coffs Harbour	2	Parkes	3
Cooma	3	*Port Macquarie*	*9*
Cootamundra	14	*Shoalhaven*	*2*
Crookwell	2	Snowy Mountains	8
Dubbo	3	Tamworth	18
Forbes	8	Taree	6
Forster	*2*	Tumut	11
Gloucester	2	*Tweed Heads*	*2*
Grafton	6	Upper Lachlan	3
Grenfell	5	Wagga Wagga	2
Gunnedah	2	Wauchope	3
Hunter Valley	4	*Yamba*	*3*
Inverell	6	Yass	7
Manning Shire	*2*		

Note: Respondents who answered with specific areas and towns (highest responses only are shown in alphabetical order, coastal towns are shown in italics).

A number of conclusions can quickly be drawn. Firstly, most people were interested in moving northwards, and there was no necessary reason for stopping at the NSW border. Secondly, bigger towns with larger, more diverse stalls and more staff, which usually meant more detailed and comprehensive information, were perceived most positively. Smaller towns with smaller exhibitions, some like Warialda virtually unknown in Sydney, were perceived less favourably. Visitors already knew something about or had at least heard of the larger towns, and size implied services. Having a well-known festival, as at Tamworth, was a major advantage in 'brand recognition'. Thirdly, many people wished to move to places that were not represented at CW, notably large inland centres such as Bathurst, Mudgee, the Southern Highlands and Orange, and every part of the north and south coast. Fourthly, there was a distinct preference for places that were not too far from Sydney; at every NSW Expo Oberon has been the nearest place to be represented. Distance was a real disadvantage, especially for small towns. People were interested in moving 'out of town but not out of touch', a scenario fitting a 'commuter countryside' perception. On the other hand, towns that might otherwise seem relatively disadvantaged, such as Moree, but mounted impressive displays (voted the best in both the 2006 and 2007 Expos) generated considerable interest, but that interest was not necessarily translated into action.

Many potential movers had pre-existing ideas about where they wanted to move and what qualities they were looking for, views shaped by factors other than CW. Housing and employment were the main factors that potential migrants sought outside Sydney, and Sydney's expensive housing market was a particular incentive to migration. Unsurprisingly these were combined with a pleasant lifestyle, often seen as being linked to coastal residence, and reasonable amenities and services. While most visitors mentioned quality of life and community as the key attractions to a potential move, this was underpinned by the primary necessity for housing and usually employment. Potential migrants were more discouraged by the costs of urban residence, the main factor that had drawn many to the Expo, rather than being lured by some notion of country lifestyles.

Queensland visitors had very similar perceptions, rationales for mobility and preferences for the coast, though they often expressed a willingness to move greater distances for good jobs. Proximity to Brisbane was still important, but just far enough away to have a sense of country lifestyle, and vague notions of lifestyle were more frequently expressed in a smaller city where housing prices and congestion were less problematic than in Sydney. What potential movers wanted can now be examined in more detail.

The Lure of the Known

While 'getting out of Sydney' or Brisbane was a pervasive theme, most did not want to 'get out' very far, and 'very far' was usually identified as being beyond a two or three hour car journey. Ironically, precisely because of this, almost all the councils at CW are beyond that distance, while visitors are more familiar with places that are closer (such as Bathurst and especially the Southern Highlands) that do not attend the Expo, until the Southern Highlands broke that tradition in 2009. Toowoomba, with a population of 90,000 and 130 kilometres west of Brisbane, proved particularly popular at the 2007 Expo. These areas had fewer problems of labour shortages and no great difficulty in gaining new residents.

Vague notions of distance as measured in time were common: 'within two hours from Sydney', 'Anywhere three hours from Sydney', 'As long as it's not more than five hours from the city' and 'proximity to Sydney'. Many visitors had little sense of what kinds of places lay within these time zones but simply knew that they had to 'get away from the rat race' but retain some ties to the city. Time was a surrogate for distance, both physical and emotional, necessitating being 'far enough' from the city to achieve a country lifestyle, but 'not too far'. A notional three hour limit was a function of many things. Visitors were much more likely to be familiar with the usually larger towns that are not too distant from Sydney and Brisbane, they were fearful of moving too far from relatives (who they might visit or would visit them), and knew little about more distant places.

For some two hours away was long enough. 'We're happy to go to the Blue Mountains, but we have friends here and the older you get the less likely it is to find friends unless you go to a church and we can't join a church for that reason', 'We don't want to go further than the Southern Highlands; we don't want to be isolated'. Others had previously tried to move over much shorter distances: 'We started looking 70 kilometres out, but now we are at the Southern Highlands and we may have to go further to find an affordable place, but the distance and the road to Oberon put us off'. For another: 'Our comfort zone is three hours from Sydney – perhaps Mudgee – we wanted to stay by the sea but sea change is more expensive than tree change'.

Town size was in itself important: one visitor to the NSW Expo, confronted by the two stalls for Oberon and Inverell that faced the entrance, in a resigned and cynical tone simply commented 'Inverell and Oberon – oh wow'. Larger places held out the obvious attractions of 'better access to shops', 'jobs for both of us' and a high school, while smaller more remote places found it particularly difficult to gain 'brand recognition'.

In Queensland notions of accessibility were somewhat different from those in NSW, both in the themes of the stalls, where some perceived air transport as commonplace, and in the aspirations of some potential migrants. As one Brisbane resident noted, speaking of Cairns which is nearly 1,400 kilometres north by air, and about 1,700 kilometres north by the most direct road:

> I'm seriously considering Cairns if the right opportunity pops up. It has an international airport, has a hospital, shopping … pubs and Brisbane is only a couple of hours away by flight. It is pretty central (quoted in Tsioutis 2007: 71).

However most Queenslanders were like those south of the border: 'We like Dalby because it isn't too far from Brisbane'.

Places that were significantly beyond the notional three hour limit experienced problems. In Inverell, 'We know that many have a three hour limit, so once they see where we are on the map we have a terrible time'. One visitor who had been to Inverell said: 'visited it, liked it, loved it, but it's too far away'. A two or three hour radius was repeatedly stated: 'no more than three hours since I have family in Sydney'. There were rare concessions to greater distance: 'five hours would be OK – that's a day's drive'. Tenterfield switched from the Sydney Expo to the Brisbane Expo for the same reason (Chapter 4). Queensland towns were again more optimistic about distance. In Roma 'our biggest challenge is that people get scared when they find out how far away we are. But you can do six hours in a day and there are towns all along the way' (quoted in Tsioutis 2007: 90). Nonetheless, as the Mayor of Quilpie, much further west, said: 'We have a sealed road to Brisbane, we just don't say how long it is. We are thousands of kilometres outside people's comfort zone'. But some Queensland visitors felt they would cope: 'any place with an airport – I still have family here'.

A majority of visitors sought a sea change and were frustrated that the Expo sub-title 'The Tree and Sea Change Expo' was not reflected in the presence of more coastal councils: 'too many trees and not enough sea'. At most Sydney Expos much the most sought after area was the mid-North Coast, about four hours north of Sydney. Still others recognised that the relatively high cost of coastal housing precluded a move to the coast. At least 20 percent of visitors to the Sydney Expo, and only slightly fewer in Brisbane, sought only a coastal move. The allure of the coast was obvious: 'I have been in Sydney all my life – moving to the country might be boring – so we need the beach for recreation'. It is unsurprising that for most residents of coastal cities images of the coast were more familiar than images of the inland.

The coast was also associated with vitality and progress, and more central compared with inland Australia: 'Central New South Wales is too far from everything'. The Great Dividing Range retained a divisive element. Unlike those who wished to move to the coast, those who were most intent on moving inland often had some form of upbringing in the country. Even then anticipations of the country could be quite vague; a builder sought to move from Sydney to Cowra because he 'had been there 50 years ago', while Katrina, who had spent some time in the large town of Dubbo, felt 'a country girl at heart' (quoted in Tsioutis 2007: 82). Very few perceived any particular lifestyles that might be superior inland, though Andrew, a 25 year old who had previously spent time inland and owned horses, declared 'I'm not a coastal person' and sought to move to either Tamworth or Armidale where there was both room for horses and a large enough town (Tsioutis 2007: 82). Such specific interests were unusual.

Visitors rarely had either a detailed conceptualisation of rural and regional life or any ability to locate particular places on the map. 'Failures of geography' were commonplace. Numerous visitors were surprised to learn about places they had never heard of and just as many were surprised to learn of facilities and resources that existed in places where they had expected little. In that very basic sense the Expos were a powerful educational tool.

> The Expo helped me to consider Quilpie and Roma. I didn't even know anything about these places before. I'm seriously considering working in the mines even. I'm now aware of other job opportunities out there.

Others had similar new perspectives. 'I will now make an informed decision about towns I had no idea about', 'A unique and informative opportunity to see other alternatives to our Brisbane lifestyle', 'I learnt a lot about places I've never heard of but are in my backyard' (quoted in Tsioutis 2007: 84, 91), 'I can now put places on the map'. Other places were elusive: 'We liked Temora [350 kilometres inland] since we want to stay a couple of hours away on the coast' and 'Glen Innes sounded good as we wanted to be no more than three hours from Sydney'. Yet to go from vivid but preliminary geography lessons to relocation required much more.

Top Places

Emphasising the preferences at the 2006 NSW Expo (Table 6.1) the towns that were most popular at both Expos in the following year (Table 6.2) were not far from either Sydney or Brisbane, were reasonably close to other large cities, were relatively large,

Table 6.2   The most frequently sought-after places, 2007

Town	Population	Distance to nearest large city
Queensland Expo		
1. Dalby	11,000	220km to Brisbane
2. Warwick	11,000	130km to Brisbane
3. Toowoomba	90,200	130km to Brisbane
4. Innisfail	9,000	100km to Cairns
5. Roma	6,800	480km to Brisbane
6. Cairns	128,000	1,720km to Brisbane
NSW Expo		
1. Taree	45,000	320km to Sydney
2. Armidale	21,600	530km to Sydney
3. Tamworth	42,500	420km to Sydney
4. Oberon	2,700	180km to Sydney
5. Wagga Wagga	44,272	232km to Canberra
6. Yass	9,700	40km to Canberra

Source: Modified from original work by Tsioutis 2007: 87.

and/or were coastal. The smallest was Oberon, the town that was closest to Sydney, while Roma was a distinct anomaly – as a small inland town. Conversely, and equally anomalous, the national capital, Canberra, regularly appeared at the NSW Expo but attracted no great interest. Its bureaucratic image had preceded it.

The more popular towns usually had populations of at least 25,000. Size was perceived to reflect a variety of work opportunities, adequate health and educational facilities and recreational possibilities, while still offering a country atmosphere. However size also meant that they were more likely to be known to some at least, and to have additional resources and diverse experts to provide a convincing display. Sharon sought to move to Armidale because of the University there;

> I need to be near shops so I can have fresh fruit and vegetables that I can buy fresh every day .. I want to have a medium sized dog to take for a walk so I can get fitter and healthier, but I won't go to the country if I can't have broadband. If you can have access to information then your brain works (quoted in Tsioutis 2007: 88).

Families sought a range of facilities: 'We want to get out of the rat race ... but we need good pay, day care, education and healthy care. We want a middle sized regional area so we can have access to lots of services. We want a social life for the kids' (quoted in Tsioutis 2007: 89). As one small town councillor recognised with some dismay: 'People from the city are probably looking for significant services as they're looking for a transition – it would be really hard to go straight to a tiny town' (quoted in Tsioutis 2007: 88). Smaller places exerted greater attractions for those who had previously lived in the country, or contemplated retirement. Those who wanted to get back to 'the way things were' also tended to prefer smaller towns such as Grenfell, that also marketed itself as 'the way things used to be'.

Why Move?

The range of considerations for moving was considerable and, in many cases, different from what inland councils initially anticipated, yet the rationale for movement was often somewhat predictable and unsurprising. Couples with children placed particular importance on the country as being a 'better place to raise kids', younger adults wanted jobs and a social life while retirees saw the country as more peaceful. Within those commonalities specific themes emerged.

Community

The perception of a superior community life in the country was widely shared in both Brisbane and Sydney, and set against a feeling of loss of community in the city: 'I want to live in a close-knit community', 'Country people have the right values', 'A slower lifestyle', 'A closer friendlier community, looking out for each other', 'A place where people would know you and say G'day in the street'. Many visitors were influenced by friendly staff on the stalls: 'The man from Gwydir spoke to us for 25 minutes. He was so genuine – you can tell he just enjoys life. We definitely will have to visit'; 'The Mayor from Dalby gave me his number so we could meet up for a coffee when I come to town', 'Grenfell had their mayor here – that shows how enthusiastic they were about welcoming people'. Many towns deliberately sent their mayors to emphasise the small scale interactions that were possible in the country.

Single people particularly attached considerable importance to community: 'I live alone', 'I am a sole parent'. Filip, a Croatian man without family in Australia, was anxious to leave for the country, feeling that Sydney was too impersonal and it was hard to make friends and meet women. Jane, a single mother of an eight and ten year old, wanted to move toward the community she felt would be there, but was concerned about leaving everyone she knew in Sydney to establish a new social network (Tsioutis 2007: 64-5). Susan had even greater concerns about moving to the country:

> I'm concerned about being a single woman moving to a country town. It was expected but disappointing to see which stalls/regions ignored me, and chose to approach families and retirees instead. This helped me decide which regions that I will research further.

Sharon, a 55-year old divorcee, also felt that councils were uninterested in single people:

> For some-one who lives alone you need social interaction. Researchers don't take account of divorced or single people. All the advertisements have elderly couples or families – but there are other people to consider. Single people are not catered for because being single has a negative image attached to it.

Both these women chose larger towns, like Armidale, as possible destinations because they felt that size offered diversity and acceptance (Tsioutis 2007: 65). For others, expectations were quickly shattered as singles perceived they were not particularly welcome.

Lifestyle

Visitors to the Expos constantly reiterated variants on the theme of lifestyle, some undeniably influenced by the dominance of lifestyle in slogans, sales pitches and most advertising (Chapter 4). Generally older people, and couples without children, attached more significance to lifestyle than younger visitors who favoured employment and housing, but anticipations of superior lifestyles were well nigh universal. Older people placed more emphasis on 'peace and quiet', 'a slower place of life' and a 'better climate' than younger people who were more likely to see a superior lifestyle as one free of financial stress.

For younger households affordability was the key to a more congenial lifestyle: 'We want to move out to a place we can afford. We're constantly paying bills. I don't want a plasma TV, I just want to be comfortable and have a place to call my own', 'I want to be able to get up on a Saturday morning and not have to worry about going to work; I want to be able to take the kids and just go somewhere'. In the city 'You have to pay to get anything'. Some sought change simply for change itself. A single mother from western Sydney observed: 'I reinvent myself very ten years. I'm a baby boomer who's wondering what to do for the next 30 years. You have done what you've had to do. The day will come when you don't have to please anyone but yourself. Your parents are dead, and your child is grown up' (quoted in Tsioutis 2007: 66). Others had a vague notion that city life was no longer attractive and the grass must be greener elsewhere: 'the city just doesn't work for me anymore'. For Danielle 'I just want to go somewhere that is remote and rural. That's my only criteria' (quoted in Tsioutis 2007: 71) – a mark of frustration as much as a real choice. At least the country sounded different. Vague senses of the need to move on were balanced by equally vague notions about the country: 'a move to the country is about stresslessness, peace, community, space ... lifestyle', 'more space and freedom' or 'a place of space and open-air – I hope it's out there'.

Climate was important for a minority of visitors: 'We want to move to Glen Innes, we really like having the four seasons and there is no drought'. The scenery also had its attractions: 'Living in Parramatta [a western suburb of Sydney] you just look at buildings all day', 'I want a large block of land with fruit trees' (quoted in Tsioutis 2007: 67). Only exceptionally were 'fruit trees' and 'acreage' of any great concern, though one Sydney resident stated: 'I need a little stream with a green field nearby where I can just sit'. Time was of value, especially the time wasted on commuting, that eroded time available for relationships. Young couples perceived this most acutely: 'You have no life. You work, you come home, then you do it again', 'I spend two hours a day on the train – it adds up', 'I have calculated that I spend 520 hours a year stuck in traffic' and 'We're sick of missing each other with commuting times. My husband leaves early in the morning and I come home late at night. We never really see each other. We really need to spend time together. The children are still young and we have the flexibility to move' (quoted in Tsioutis 2007: 68). The country seemed to offer more time, the opportunity for recreation, exercise and superior health, and affordability. Many visitors had at least absorbed the council messages.

Employment

Because unemployment levels have been very low in Australia, employment was not a key stimulus to migration, although unemployment was relatively high in parts of western Sydney and Brisbane. Indeed the very high employment levels in much of the country, especially in the booming mining areas of Queensland, created the need for CW Expos as regions sought to fill job vacancies (Chapter 3). Employment was nonetheless necessary for most of those who wished to move, especially younger people; most assumed that it would be available in the country, and the councils emphasised this was so. Others interested in moving had come to the Expos primarily to find out which towns had vacancies for their particular skill, and start from there in finalising a destination. Some knew that their skills, for example in nursing, were universally required but sought opportunities for their partners.

Understandably, people were unwilling to move without reasonable job prospects: 'job security would make the move easier', 'We'd move earlier if we got offered a really good paying job', 'All country jobs should be advertised in city papers. Lots of people cannot move from a job to a town unless they are job guaranteed. This affects my family'. A caretaker in his 50s stated 'I'm considering a move to Dalby because they were offering me a job on the spot and allowance money' (quoted in Tsioutis 2007: 73). He was one of the extremely rare examples of visitors to the Expos who seemed likely to move after one brief Expo experience.

Some visitors were willing to change occupations simply to be able to move to the country, notably the very few who wanted to change to 'work on the land' and the rather larger number who sought to establish small businesses. Such people were more circumspect over what might be possible. A Brisbane couple, both of whom were retail managers, sought to move out of the city, embark on a new lifestyle and set up a café somewhere in northern Queensland. The Expo was their opportunity to assess where they might go that had relatively few cafés and where there was an adequate population base for a new one. Concern over population size and characteristics was a major theme for others who sought to transfer or establish businesses. As one individual, who managed to combine being both a panel beater and a physiotherapist, noted

> I don't want to work hard just to survive any more. I heard it was cheaper to live
> in regional areas. My only problem is that I need a regional centre, not a quaint
> little town because I need business and people so I can make a living (quoted in
> Tsioutis 2007: 75).

For some, jobs were everything: 'I just want good opportunities – the place is not important'. A number of tradespeople and others were willing to move to the country because jobs were available there, whereas they had experienced difficulty finding good employment with adequate wages in the city. Several such visitors were relatively new migrants to Australia, notably from India. At both the 2007 Sydney and Brisbane Expos migrant doctors from Africa and Pakistan were also seeking work in the country. Such recent immigrants tended to be more flexible in their ability to move, having developed few strong social ties in the city and were living in rented accommodation, but none were well informed about country life.

Teachers were well represented at all Expos. Many young teachers were willing to take rural placements in the short term since, in the regulated public education job market, that would very significantly improve their chances of a good urban position later on.

> I cannot get a job in the city or east of Parramatta [just west of the geographical centre of Sydney]. It is easier to get a job as a teacher in Western Sydney but the students are rough and they are hard to control (so are the parents). We really have no choice but to leave Sydney in order to make some money, save up and move somewhere else. I'm thinking of working at Lightning Ridge [a tiny opal mining township close to the Queensland border] for three years; it is a rough place but with the system in place right now, I will get priority over where I can choose a job next time. I want to move to a better environment to raise children (quoted in Tsioutis 2007: 73).

Particularly in the context of employment not all decisions to move to the country were entirely voluntary, but were hedged in by restrictive employment conditions in some areas and by assumptions of quite imminent return. While a regulated system takes many teachers to country areas for extended periods, many simply returned to the city after a few years – a transient situation that can be disruptive to regional education.

Housing

Housing was a major concern for both income and lifestyle reasons. Affordability was of great concern for younger people, and singles, but with increased age greater emphasis was placed on particular kinds of housing, and often with the possibility of some land. Many younger people, especially those with children, found the price of adequate urban accommodation simply out of reach. One couple with two young children epitomised this situation: 'despite family being here, we have to move … we just cannot afford Sydney, we have no savings'. A pregnant mother asked rhetorically: 'Do I want my kids to grow up in Sydney? In our little apartment? We pay so much for a tiny little place. We want to enjoy our lives and have a better lifestyle'. Housing costs were constant issues: 'house prices are forcing us out', 'we can no longer afford to live in Sydney', 'we rent; we can't afford housing',

The housing choices of older people were closer to being lifestyle decisions. A 60-year-old couple nearing retirement wanted to 'buy a big house with land. House prices in Cowra allow us to afford what we want' (quoted in Tsioutis 2007: 70). Some, however idealistically, had particular requirements. One single woman stressed:

> I need a minimum of 70 square metres … I want a spare room for visitors, because I live alone. I also want land. I want something to look at, as in distance … I want to be able to look out into the distance and not have anything blocking

my view. I would stay in Sydney if I could have what I want, but I can't afford
the places that I would want to live, for example, the Northern Beaches. I don't
want to live there if I have to live in a unit and look at apartments' (quoted in
Tsioutis 2007: 70).

For many, housing and the space that went with it, was the most critical component
of lifestyle.

Services

None sought to leave the cities because of any lack of services, though affordability
was an issue. Most were in some way concerned with service provision, for
themselves and their children, and whether appropriate access to services would be
possible in smaller towns. The single most important service was health care, though
education was often stressed. Councils that were large enough to be able to offer and
display a diversity of education facilities, such as Armidale, were praised.

Older people, invariably more likely to be conservative about migration and
leaving the familiar, were particularly concerned. As one 75-year old man said:

I would love to go but at our age we have to think of services – where are the
hospitals? Some are miles away, and you would need a change of horses, and I'm
not allowed to drive at my age.

Even lengthy discussion with councils could not always disturb reasonable fears that
smaller places would have fewer and more basic services. However few if any visitors
actually realised that the presence of State Education and Health Departments, and
the Rural Doctors Network, primarily to encourage health workers and teachers
to move to regional areas where plenty of vacancies existed, alongside other state
departments with similar missions, indicated a shortage of workers in key services
in the regions.

When asked what they might miss about the city if they moved to the country,
most were convinced they would miss very little. Those who had previously lived in
the country perceived some disadvantages. For one middle aged couple, the husband,
who had never lived in the country, said he would miss nothing, whereas his wife
felt she would miss shopping and other amenities. Many did expect they would miss
family and friends: 'I will miss my friends. I'm not worried about convenience,
all the basic things that you need are available', 'I will miss everything at your
fingertips … my family and my friends. I'm worried about making new friends in
the community and fitting back into a community' (quoted in Tsioutis 2007: 71).
Destinations were particularly attractive if friends and relatives were already there:
'I am probably more likely to go to Cairns because my family is in Cairns'.

Security, Congestion and Lost Community

Although inland councils and CW staff believed that security was a problem in capital cities, and some who had moved claimed that security had been an issue (Chapter 8). As often as not, like the amorphous 'lifestyle', this moral panic was expected of them. Nonetheless, partly fuelled by nostalgia, many believed that urban life had changed for the worse, with greater congestion, pollution and the loss of neighbourliness, and that smaller towns would be rather more like a community, with friendly neighbours. Such perceptions were more forcefully expressed in Sydney where some visitors conveyed a sense close to panic: 'I hate Sydney', 'Anywhere but Sydney', 'I'm sick of Sydney' were frequent responses, though not often or easily elaborated upon. Urban expansion and redevelopment were constant threats: 'If Penrith starts to look like Parramatta in a few years time then we'll consider moving'. Some councils attracted interest from people who wanted to move from the NSW coast, since it had become 'too much like Sydney'. Some visitors had previously moved to avoid such congestion. One Brisbane couple in their 50s had moved from the suburb of Merrylands in western Sydney to Brisbane 23 years earlier:

> There were too many people in Sydney; it was grotty, the public transport system was bad and we had a big commute. That's why we moved to Brisbane. Back then is was like a big country town. But now there is too much traffic in Brisbane, there are too may people and there is no water.

They claimed to be intending to move to Thursday Island, a small, largely Melanesian island off the far north coast of Queensland (Tsioutis 2007: 78). Traffic was invariably a problem: 'I don't believe in centralisation, cities are growing too much, we're sick of Sydney's traffic, crime and attitudes of people' (quoted in Tsioutis 2007: 76). Such largely inchoate fears were common. 'We want a greater freedom for children, where they can run and play and ride their bikes just as we used to do without fear of traffic or paedophiles'.

Some fears were however well grounded: 'We've been robbed three times and now we have to get shutters – it's like being in a prison. I want to live in a neighbourhood where you can feel safe to walk your dog'; 'We had a drive-by shooting in our street the other week. Enough is enough', 'I have 13 locks to get into my house; I would just love to get back to a situation where security is not an issue'. For a young man who wanted to move to Tamworth crime was the catalyst: 'I have become sick of commuting and just recently my car was stolen' (quoted in Tsioutis 2007: 77).

The apparent loss of community and neighbourliness was a constant refrain, with numerous people referring to the need to 'get back' to a time or place where a sense of community still existed. The past was a more amicable country, as perceptions of a dystopian city had their converse in romanticised notions of country life. But the main disadvantages of urban life were not any loss of community – many had relatives there – but the costs of commuting and time wasted. A familiar refrain was 'We want to get out of the rat race to a simpler life'. Many sought to bring their children up in a different environment: 'we want them to ride their bikes – to do what we would do as kids – they lose their innocence too early in Sydney' where 'there is space to play and run and not worry about crime'. Quite simply the increasingly 'built-up urban environment' seemed to impose too many restrictions and problems.

White Flight

For a minority moving away from cities that appeared increasingly multicultural and not those of their youth was of some importance, and probably of greater significance than expressed in personal survey data and conversations. Like large cities the world over, the social composition of Sydney and Brisbane has changed considerably in the past quarter of a century, with migration from non-English-speaking countries. While the proportion of Australia's population that was born overseas at the 2006 Census remained steady at 22 percent (the same as it was in 1996), the total number of overseas born Australians rose to 4.4 million, and recent arrivals were disproportionately from non-European countries, especially in Asia. Most migrants tended to concentrate in the larger cities, and in particular parts of them, with Sydney having 32 percent of its residents born overseas and Brisbane 22 percent. In contrast NSW had 24 percent of residents born overseas and Queensland only 18 percent; few overseas-born lived outside the capital cities. Many small towns, such as Glen Innes and Oberon (Chapter 8), had almost no international migrants.

Racial concerns were more frequently expressed in Sydney. As one resident of the suburb of Denistone noted: 'Sydney lacks a community solidity, neighbourly friendship, zero community spirit; increased crime, Asianisation of previously European origin areas. We are not racist but it is unsafe to walk on streets in the vicinity of my area at night'. Another visitor observed 'I do not enjoy raising my son and daughter with Muslims' and another believed that Armidale might offer his solution: 'I want to move out of Sydney in order to get back to an adequate ethnic balance'. Another came close to tears as she stated:

> I'm not racist but … why do these ethnics get more benefits than us? My grandfather fought in World War Two, we've been paying taxes for years like good, honest citizens and these people who come and have all these benefits that we cannot even enjoy. There is an ethnic injustice. Our whole suburb has become Asian and now we feel like the minority. You walk down the street and you can't understand the shop signs. [The suburb] Epping has become Eiping and Eastwood has become Eastwoo.

When it was suggested that the same situation might also apply in the country, she responded with greater frustration: 'Yeah but there's fewer of them' (quoted in Tsioutis 2007: 77-8), exemplifying what has been described for Britain as 'the countryside as a place of white safety' (Holloway 2006: 8, Neal 2009). Some had merely 'had enough of the riff-raff', a marginally more subtle form of distaste.

An attraction of sea change for many has long been that coastal Australian towns are predominantly white, or perceived so compared with Sydney. Mainly anecdotal information on the coastal town of Port Macquarie indicates that for some older people especially, residential satisfaction was linked to the town's homogeneity – a place predominantly of white people and especially 'no Asians' (Kijas 2002). Metropolitan transformation has hastened some forms of migration:

> The people down the RSL [Returned Service League Club] do go up the coast. They don't feel at home anymore. It is a different type of place, what the local shops stock, who they meet at the pub ... for all sorts of reasons and, without wishing to sound racist, many friends are selling out and moving to the North Coast. They don't feel Fairfield is the same. The reasons are endless, lifestyle related. They hate the traffic and big city life but it's also about how they feel about the changes in the neighbourhood (quoted in Morris 1999: 7).

Ironically many relatively recent migrants to Australia from Asia have found achieving economic success in capital cities relatively difficult and ambitions of house purchase, and other goals, have been put on hold. Even over the short six-year period since the inception of Expos there was a significant increase in the number of 'new Australians' seeking to move to regional areas. Few knew much about regional areas, although some were cautious about areas that they perceived as relatively 'white' and where there might be an uncertain reception. A Chinese couple queried: 'we think country people are very friendly – but would we fit in with a more redneck population?'

Some councils at least professed to be concerned about the existence of white flight, and the kinds of people that might bring. As Cowra pointed out 'these are not the people we want to come to our community; we welcome a slow flow of others to diversify the community'. Councils were rather more conscious of the need for change than visitors. With a lone exception in Sydney no visitors talked about Aboriginal issues, which for most urban residents are quite other worldly. The idea of safety in whiteness, and the apparent absence of Aborigines from stalls and from most promotional material, raises questions about the nature and ownership of the country.

Staying Put

Many visitors to the Expos will probably never move to the country, however much they proclaimed this to be their objective; they were there not as considered future migrants but for a vicarious visit to the countryside. Not surprisingly many of their responses were fickle and superficial, but an opportunity for inchoate rage against urban life. Many more, seriously interested in moving, were simply bewildered and mesmerised by the choices available. Others realised that a hard decision had to be made and they were not at that point: 'Just taking the "big step" is the only issue for me', 'Sometimes you just have to jump … and a net will appear, you have got to make a decision sometime. We just had to make the move; you're never going to be 100 percent ready' (quoted in Tsioutis 2007: 92). Inertia reins back many, and moving from the familiar is a step too far for some.

Family ties were repeated brakes on mobility. Most people do not move alone but are involved in family decisions, and families have to cater for all tastes and interests, which may be better served in large urban centres rather than in unfamiliar small towns. Couples that were about to start a family, or with very young children, were more likely to move than those who had children in school. Similar marketing strategies in Sweden specifically targeted households where the oldest child had not yet started school (Niedomysl 2007). If the children were teenagers, perhaps fractious in themselves, moving was even more difficult since 'they might not want to leave their friends'. Numerous families were simply waiting for their children to leave home: 'We can't tell him this but we're waiting for our son to move so we can move', 'We're only staying in Brisbane because of our 20-year-old daughter. She wouldn't move with us because she is madly in love with a boy in Brisbane' (quoted in Tsioutis 2007: 79).

Several households had divided loyalties, where one partner wanted to move and the other was reluctant. Frequently, if one partner had come from the country, it was that one who wanted to return and the city-born partner who was not interested. In this reluctance there were no obvious gender divides, though women were more likely to emphasise what they might miss – mainly shops and relatives. One man sought to move to inland NSW but his wife was reluctant to move inland hence they had compromised by choosing a coastal location. Such compromises were frequent. Though divorce was light-heartedly suggested, stability was the more probable outcome.

A Sense of Place

Remarkably few visitors had a clear sense of where they wanted to go beforehand; as they left the Expos that had rarely been clarified. Many had a sense of distance rather than a particular region, but had been bemused by the variety of places on offer. The Expos offered a 'sampling plate' that tended to inundate visitors with information rather than clarify initial ideas and perceptions. At the end of the Expos some 68 percent of Brisbane visitors and 56 percent of Sydney visitors were undecided even on which towns they were particularly interested in. Very few had narrowed the choice down to one or two.

Many visitors sought coastal locations, but recognised that sea change was invariably more expensive and that an inland tree change would have to be, for them, a second best outcome. Many potential movers consequently went through a series of 'stages', loosely from 'interested in learning more' through 'considering moving' to 'have decided to move'. In some cases such stages coincided with successive Expos. Several were visiting for the second or third time, supplemented by visits to towns. More than 10 percent of visitors to the 2007 Sydney Expo had been before. A handful had made the decision to move but came to the next available Expo to reassure themselves that they had done the right thing and chosen the right place. One visitor said: 'I found each stall that I went to very interesting and informative. I found the people helpful and encouraging about the different resources found in the areas they live in. I will be back next year'. Another stated that 'there will be questions in the future. I first have to read the brochures and contact relevant councils and real estate agents. I estimate this will take me through to the end of September'. As one visitor noted about the exhibitors: 'It is really hard to choose between them. I have to visit, they're all so positive. They're not going to tell you about the abattoir or the high crime rate' (quoted in Tsioutis 2007: 91). Abattoirs were something of a bête noir. Many visitors left the Expos realising they needed further information – notably on the specifics of house prices and job vacancies – and at the very least that they needed to travel to regional Australia and examine the reality, get on to the internet, or return to the next Expo. For others timing was as important as taking time. Some were waiting for retirement, for children to leave home or for a job vacancy. That might mean particular windows of opportunity: 'We have to move very soon before the first of our four children gets to high school age'.

At the 2008 Sydney Expo a sample of 80 visitors were specifically asked if they had visited the Glen Innes and Oberon stalls. Half had missed them or had no recollection of them, partly because they had other specific objectives. Others had simply grabbed the showbags. Almost all those who had thought about Glen Innes regarded it as too distant: 'too far', 'too far inland', 'the wrong geographic direction', 'a long way from anywhere' and 'too far from the coast'. That was combined with a perception that it was 'too cold', 'minus 7 degrees', 'too cool for me. I went there once, stopped, got out of the car, got straight back in again and drove on'. Oberon too was similarly seen by many as too cold, and too small for adequate services or adequate job opportunities: 'a bit small, we don't want to give up quality of life and income'. Such negativity, much of which came from familiarity, dominated more vague positive statements: 'they seemed nice', 'Glen Innes has always been appealing', 'they went out of their way to help us'. Just one visitor had a specific positive comment: 'sapphires'. Both Oberon and Glen Innes had stalls that were relatively plain and less memorable than others at the Expo, hence they struggled to gain immediate interest. Most people needed much more time and familiarity to assess possibilities and make informed decisions, and neither Glen Innes nor Oberon seemed as attractive as larger, warmer towns (or closer towns in the case of Glen Innes). In a competitive world, and a large Expo, first impressions counted: 'public attention is the scarcest of commodities, and … to be of consequence is to be seen and heard' (Bauman 1995: 153). Other small towns had similar struggles.

The two most important factors that influenced visitors were size and location – large enough for a reasonable range of services and not too far from present locations: urban conveniences without city hassles, and accessible – the best of both worlds, perhaps hybrid country. Families with children were the most demanding, seeking a range of facilities to meet their diverse and multiple needs, and thus the most difficult to attract to the smaller places that needed them most.

While most visitors were pragmatic in their necessary focus on housing and employment, and an adequate range of accessible services, they also sought a reasonable quality of life in a pleasant environment. It was couched by many as the opportunity to live with fewer financial stresses, especially housing costs, and more time for family, rather than notions of acres of fruit trees or greater self-sufficiency in wide open, unpolluted spaces. Although this generally coincided with what the councils were offering (Chapter 4), and councils had moulded their presentations to meet such notions, limited understanding of regional life meant that perceptions were often clichés rather than the outcome of experience and knowledge. Attitudes of visitors had often settled on polar opposites: the open spaces of the country compared with the apartment blocks of the city, the crime ridden suburbs compared with the concerned watchful eye of a vigilant tightly knit rural community, or the abrasive city people and the friendly, relaxed rural community. In some respects such a dichotomy was exactly that posited by CW and by a host of council spruikers. This rural–urban division was often tinged with nostalgia – a feeling that movers might find greater security, community and even a more monocultural community – but at least a place where scale was more manageable, and there was a feeling of stability alongside a sense of open space. To an extent this nostalgia applied to the councils, since attracting new migrants could bring skills and businesses and a return to more prosperous earlier times.

Pragmatism was most evident amongst those who were more serious about moving – and lifestyle considerations weighed less heavily with them– but aspirations and intent had yet to be tested. The following chapter therefore examines who actually moved to regional Australia, using case studies of two small NSW towns – Oberon and Glen Innes – to examine aspirations and intentions, the challenges and consequences of relocation, and the extent to which there were disjunctures between what councils were advertising and branding, what Expo visitors were looking for and what new migrants experienced.

Chapter 7

Taking to the Country

People do move to regional Australia, not necessarily from metropolises and irrespective of Country Week (CW), and migration has occurred for many reasons. Even the smallest towns that struggle at the Expos have experienced in-migration, albeit counteracted by losses, and but it is the larger regional centres and coastal towns that generally have grown fastest. This chapter examines contemporary migration into the two small towns of Oberon and Glen Innes, much like others in regional Australia, if not the larger towns. Neither generally attracted great interest at the Expos (Chapter 6). The subsequent chapter looks in more detail at some of these new arrivals, on a broader regional scale, as portrayed in CW documents and in the media, to examine what common and consistent themes are present, and how images may diverge from reality.

Oberon and Glen Innes

Oberon and Glen Innes are relatively small towns in rural NSW, but reasonably typical of many towns smaller than 'sponge cities'. Oberon is 200 kilometres west of Sydney and Glen Innes is about 800 kilometres north of it, and closer to Brisbane. Both are in upland areas, more than 1,000 metres above sea level, with distinctly cold winters, and both claim to have the pleasure of four seasons – whereas spring and autumn barely register in coastal areas. (Oberon Council's webpage proudly depicts a snow-covered tourist office). Oberon has a population of about 2,700, at the centre of a local government area with 5,000 people, while Glen Innes has about 5,700 people at the centre of the Glen Innes Severn council area containing 8,900 people. Both have what are effectively one street 'central business districts', low density detached housing surrounding it and a scatter of more dispersed properties on small acreage blocks beyond that.

Since the 1930s Oberon's economy has been dominated by a timber mill and population changes have partly reflected its fluctuating fortunes. By contrast Glen Innes is a more typical regional service centre with a range of small businesses linked to surrounding agricultural activities, but both towns have significant public sector employment. The urban populations have increased slowly since the 1980s. Glen Innes particularly has a population that is significantly older than the state as whole, especially in age groups over 55, largely the outcome of limited immigration and significant outmigration. Both towns have relative deficits in the age group 15-44. Glen Innes has a substantial number of visitors because of its position on the New England Highway and has sought to develop tourism through highly effective marketing of Celtic Country and sapphire mining.

Typical of much of regional Australia, the town populations are dominated by a generally long-established population of northern European ancestry with a small, usually fringe, Aboriginal population. In Oberon local government area some 2 percent of the population described themselves as of Aboriginal ancestry, and in Glen Innes the comparable proportion was 6 percent, representing numbers of 108 and 467 respectively. The proportion of people who speak a language other than English at home is just 3 percent in Oberon and 1 percent in Glen Innes (compared with 26 percent for NSW as a whole) reflecting the striking absence of international migration in much of small town Australia.

Both towns have hospitals and high schools, libraries and banks, and a choice of supermarkets. Oberon lacks any significant public transport, though it has a weekday bus service to Bathurst, a three day a week coach service to Mount Victoria for rail connections to Sydney, and a limited taxi service. Cars are the order of the day. So too in Glen Innes, where there are infrequent bus services to Armidale, Inverell and Grafton. The closest airport is at Armidale, an hour's drive away. The New England Highway gives Glen Innes more of a 'buzz' and a sense of being a central place, than smaller and more sleepy Oberon, despite its industry and relative closeness to Sydney. A range of sporting facilities exists in both towns, notably golf clubs and sports pitches for cricket and various football codes.

Glen Innes and Oberon have appeared at every CW Expo, but have used slightly different strategies. Oberon's stall has emphasised property prices – a function of its relative closeness to Sydney – whereas Glen Innes has emphasised Celtic distinctiveness and the potential for residential satisfaction and business development (Chapter 4). Both have attracted new migrants, with Oberon especially being significant for migration from Sydney. As a real estate agent stated in 2006, with no small degree of hyperbole: 'Ten to 15 years ago you could have shot a cannon down the main street of Oberon and not hit anyone. Now you would kill half of Sydney' (quoted in Brown 2006: 25). In Glen Innes new residents are from more diverse origins but in both towns there has been a steady inflow.

In both Oberon and Glen Innes interviews were conducted with local people, but primarily with random samples of residents who had moved there during the previous five years. In Oberon some 16 newcomers were interviewed in 2006, whilst two years later 30 households were interviewed in both of the towns, in all cases mainly located by snowball sampling. In 2008 this represented some 71 people in Oberon and 56 in Glen Innes. The households had been present in the towns for periods of three months to five years (Brown 2006, Cuomo 2008).

Going to the Country

New migrants to both Glen Innes and Oberon were significantly younger than the established population. In both towns newcomers were mainly between 25 and

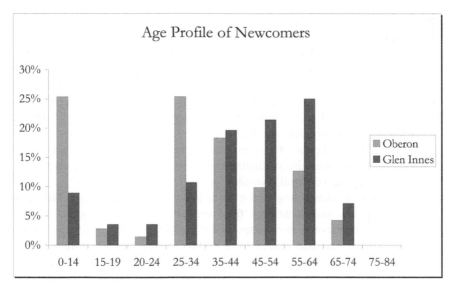

Figure 7.1   Age profile of newcomers to Oberon and Glen Innes

64, as might be expected, but those in Oberon were younger (Figure 7.1), partly reflecting employment opportunities in the timber mill, while people older than 65 were largely absent. In both towns the newcomers were mainly households, some with young children, so that some rejuvenation of the local population followed. Households in Glen Innes were somewhat older and smaller, further reflecting the greater significance of migration for employment in Oberon and for retirement to Glen Innes. In the 2006 sample three quarters of those who had moved into Oberon were married, with the remainder divorced or single. Just one was living alone and the rest were living in family contexts, and in detached houses, so that the median household size was just over three (Brown 2006). A similar situation was true in 2008 when roughly a third of all migrants were single person households, a third were couples and a third were couples with children, but the differences between the samples emphasises the diversity of migration. In Glen Innes in 2008 over half the new households (54 percent) were couples without children, and only 13 percent were couples with children. Here too a third of the migrants were single person households. Only in Glen Innes was there a lone single parent household. Single person households tended to be younger, male and had typically moved primarily for employment. All the couples were in heterosexual married relationships.

Older people were less likely to migrate because of preference for familiar surroundings, and perhaps also declining health, unless they had close relatives in the towns. However many of those who had moved before the age of 65 had already retired, especially in Glen Innes. Only a third of the adult newcomers to Glen Innes were in full-time employment, compared with more than half in

Oberon, but of those who were officially 'retired' some 25 percent were working, with some having started their own businesses.

The vast majority of new households defined themselves as 'Australian', though one household defined itself as Aboriginal. Only 13 percent of all the migrants were born outside Australia, a substantially lower proportion than in the Australian population as a whole. Most of the very few overseas-born came from Britain or New Zealand, and only two were from non-English-speaking countries: Italy and Japan. Similarly only one person claimed a religious denomination other than Christian. New migration thus constituted a continued process of 'whitening' of small town populations already distinguished from metropolitan populations by their lack of ethnic and cultural diversity.

Much new migration to regional centres is not from state capitals but often from smaller towns, and sometimes from great distances. Nonetheless Sydney dominated new migration to the two towns. Of the 16 who had moved into Oberon in the first sample nine had moved from Sydney. Three had come from nearby Bathurst (so that movement into sponge cities is actually a two-way process), three from smaller NSW towns and two from other states. Most movement was therefore over relatively short distances (Brown 2006). In the second sample some 46 percent of new households in Oberon had moved from Sydney; coincidentally, even in Glen Innes, hundreds of kilometres distant, 46 percent of newcomers had also moved from Sydney, while some 20 percent of newcomers had moved from Brisbane. In Oberon six percent of newcomers had moved from New Zealand, all for timber mill employment, a rare example of international migration directly to regional areas. Other migrants had come long distances, with some 15 percent of migrant households in both towns having come from other states (in addition to those from Brisbane). The remainder, some 37 percent in Oberon and just 17 percent in Glen Innes, had come from elsewhere in NSW. In considerable contrast none of the new households in Glen Innes had come from within two hours drive of the town, in a region where there are several other similarly sized towns. In Oberon however some 16 percent of households had moved from larger regional centres such as Bathurst and Lithgow, to buy more affordable property in the area or move to a smaller town.

Distance from their previous residence, especially from either of the state capitals, Brisbane and Sydney, was influential in decision-making. In Oberon being within a three hour drive of Sydney was a significant advantage, and many newcomers frequently made day-trips to visit families and friends, even on a weekly basis. Migrants rarely sought to utterly disengage from their previous locations: 'keeping in touch' was important. Many of those moving to Oberon found it convenient for Sydney, whether because of friends or family, part-time employment there or because the airport was reasonably accessible (Brown 2006: 96). Both in Oberon and Glen Innes, the proximity of the much larger towns of Bathurst and Armidale respectively, about an hour's drive away, were welcome since they provided access to some services unavailable in the smaller towns.

Why Move?

Most households had relocated for a variety of reasons. In Oberon these tended to be dominated by economic factors, centred on employment, and in Glen Innes by a stronger emphasis on a 'change of life' often through early retirement (Table 7.1). Economic factors involved employment, with 40 percent of Oberon households and 27 percent of Glen Innes households emphasising its primacy, but reduced housing costs were the primary influence on several households, especially in Glen Innes. The only numerically comparable proportions were in Glen Innes where 37 percent of all households moved for a 'change of life', which mainly involved retirement linked to a desire to live in the country. Motivations were interconnected; retirement provided the opportunity for a change of life that a smaller town provided.

Oberon, with its timber mill, was unusual in having one large source of employment, and thus a greater employment catchment than towns of comparable size, such as Glen Innes, where broadly based service provision dominated. A third of those who had migrated to Oberon moved because they had secured employment there, usually either in the timber industry or in such institutions as the NSW Police Force or the National Parks and Wildlife Service. Within the public service certain transfers are straightforward hence various public servants move relatively easily between similar activities – whether in education, health or correction services – with the knowledge that further transfers can be arranged. Some of those moving to Oberon had come from rather tougher urban jails, to

Table 7.1   Reasons for moving to Glen Innes and Oberon

Motivations	Oberon	Glen Innes
Economic	**40%**	**27%**
Employment	33%	10%
Housing costs	3%	13%
Costs of living	3%	3%
Environment	**23%**	**30%**
Climate	3%	10%
Live in the country	17%	20%
Connection to 'country'	3%	0%
Stage-of-life	**13%**	**37%**
'Change of ;ife'	3%	13%
Retirement	10%	23%
Social	**23%**	**7%**
Romance	17%	0%
Kinship	7%	7%

Note: Percentages may not always add up exactly due to rounding.

Source: Cuomo (2008, 57).

Oberon, where the facility was not high security, was relatively small and located 30 kilometres outside town. Lauren, a police officer, had grown up on a property outside the NSW regional centre of Dubbo. She had moved away for training, then spent four years working in Sydney before she could obtain a posting to Bathurst; a year later an opportunity came to transfer to Oberon, so she moved: 'I prefer to work in small towns because of the quietness; [Oberon] is a much nicer town than Bathurst because it is smaller and has a better lifestyle' (quoted in Cuomo 2008: 58-9).

Over 70 percent of those who had moved to Oberon mentioned house prices. Couples had often searched other areas, including coastal areas, before recognising the relative price advantages of Oberon: 'We were looking at the Southern Highlands Sunday after Sunday but we hadn't found anything ... we came up and looked at rural properties and realised that the only difference between the Southern Highlands and Oberon is $200,000' (quoted in Cuomo 2008: 59). David and Nicole, a teacher and health administrator respectively, moved to Glen Innes because of family ties, house prices and greater safety for their children. They had sought good housing in Sydney but were unable to afford it' (Cuomo 2008: 60). Just three out of 16 movers to Oberon stressed housing as a primary motivation. Despite its relative cheapness, even the cheapest housing is inadequate without employment and other services. A significant number of migrants to Oberon first purchased houses there as investment properties before eventually moving themselves (Brown 2006: 100-101), reflecting one version of the stages that are involved before definitive migration.

House prices might however take people beyond where they initially preferred. Glenda had moved from the lower Blue Mountains, on the fringe of Sydney, because her inadequate superannuation funds meant that she had to sell her house but she wanted to remain within reach of her three children, in Armidale, Oberon and Canberra, and her friends: 'I knew this area, I knew it was flippin' cold but I could not afford elsewhere. I could have bought a house in the Upper [Blue] Mountains but then I wouldn't have had much money left over. Although I can't hop on a train and visit my friends at least here I see my children all the time' (quoted in Cuomo 2008: 60). Jane had moved to Glen Innes after a series of health and financial problems had left her with no personal savings or superannuation; rents were much cheaper, she could live comfortably on a disability pension and she no longer needed a car as everything was within walking distance. Louisa moved from Brisbane to Glen Innes because she had inadequate money for anything more than an apartment there, but could now have a house with a garden and dog, after selling her half share in a house jointly owned with her mother (Cuomo 2008: 60). As such examples indicate, choices could be forced upon financially constrained migrants, and inaccessible house prices could disrupt existing social ties. Being partially forced to move to regional Australia may result in harboured resentments that discourage a successful transition.

The notion of living in the country exercised a particular attraction and could be encapsulated in the desire to move to a cooler climate (away from greater coastal heat and humidity) and live in a rural setting. Climate was a factor in two unusually temperate towns; 'The heat was the thing that was a drawback for the coast. I can't handle the heat. While here [Glen Innes] I can handle the cold and the summer is wonderful because there is no humidity' (quoted in Cuomo 2008: 63). Another couple had moved from a retirement village on the coast to Glen Innes, in the quest for a cooler climate, good facilities, a friendly community and several social groups. Conceivably cold winters discouraged more people than they attracted.

For many the desire to live in the country was also a wish to 'live on acreage', where scenery might be better experienced and appreciated, though many newcomers still lived in the towns. Some sought hobby farms, where they might grow their own food, but few achieved that ideal. Occasionally greater self-sufficiency was achieved:

> Our plan was to run some cattle or sheep but nothing too big because we are too old to be able to run a big farm the way it should be. We are also in the process of being sustainable, with solar and wind power and growing our own things. It had always been a passion but it is not realistic to achieve this without having the money to do that ... Long term we want to be self-sustainable in water, electricity ... and we want to rehabilitate the river on our property because I think we have a responsibility to look after the land (quoted in Cuomo 2008: 64)

Such idealistic aspirations were rare. Frequently more commercial considerations proved to be the catalyst. Andrew grew up on a large cattle property on the outskirts of Glen Innes, but left after completing high school and moved to Sydney to study. There he met Maria (also from the country) and they married. They started a family and, with their third child on the way, were considering whether they should get 'a Sydney mortgage'. They decided to move back and run the family's cattle breeding business: 'our main driver was quality of life, also being close to relatives and a safe place for kids' (quoted in Cuomo 2008: 64). The actual landscape could also be a factor. For Richard and Emily, an Oberon couple with four children, the country environment offered space for their children to play; as Emily observed:

> I love where I live. I love the actual property. I love the scenery, it is beautiful ... I get all romantic and sentimental about it. I was a very sentimental child and when we used to come up here, we would camp in the most glorious places. I just longed to live here and I longed to live in those places we visited when I was a child. To me it is a physical beauty thing ... that's why I wanted to be here (quoted in Cuomo 2008: 61).

Coinciding with retirement and financial independence another couple sought the country:

We wanted to go to the country. We didn't want to go to the coast because the sea doesn't interest us. The reason we picked here [Oberon] was a combination of reasons. One of the primary considerations was that it was close to the city as our family is still there. We also wanted a cooler climate because we were sick and tired of the humidity in the city ... we were looking for an area which was easy on the eye, pretty and very scenic (quoted in Cuomo 2008: 61-2)

Here too, retirement migration alongside financial stability enabled careful, long-term consideration of migration, the development of a 'check-list' of necessary factors and, literally, time to check the list. Oberon was the best possible 'compromise of all those ideals' in balancing proximity to Sydney, climate and landscape. Significantly this couple recognised that any choice was a compromise and none could ever be perfect.

The often nebulous lifestyle, or a 'change of life', was a significant influence especially in Glen Innes. Some 37 percent of newcomers to Glen Innes, evenly spread across age groups, stressed lifestyle. By contrast in Oberon only 13 percent of newcomers identified lifestyle reasons and none were under 35; most were over 56 and simply wanted to retire to a slower pace of life because they 'felt overwhelmed by the hectic pace of the city' (Cuomo 2008: 65; Brown 2006: 92, 103). Employment and affordable housing were always more important; new lifestyles hinged on adequate housing.

The idea of community was part of the attraction of migration but especially where that also constituted downsizing: 'We came here for the community. In the city people are sold community that is not real. Suburbia is isolating, it is a car-driven society. It is a dormitory. But here there is a community, people know who you are and they can recognise strangers' (quoted in Cuomo 2008: 69). While moving to a smaller place for a new lifestyle, cheaper housing or other reasons implied some dislike of circumstances elsewhere, 'push' factors were rarely mentioned. Distaste for urban life and assumptions that urban congestion and crime had reached intolerable levels were rarely evident in migrants merely seeking to make improvements to their lives. Urban dystopia might have been at the back of some minds but it was never in the foreground. Yet the alien city was a factor for some. One 'didn't like the fact that no-one cares in the city'; others were threatened by the 'rat race' and 'clogged roads and busy services'. One, who had moved from Sydney, was concerned at a stabbing and a shooting in nearby schools. But it was not only metropolises that posed concerns. Bathurst, with just over 30,000 people, was seen as 'too crazy' by one newcomer (Brown 2006: 95). Another couple moved to Oberon since Bathurst seemed 'more like Sydney'. Ultimately for most of those who emphasised lifestyle it could simply be summarised as a quieter, less stressful life usually facilitated by retirement and migration to a smaller community.

Social factors could be important. Romantic attachments brought individuals to join partners and many newcomers moved closer to kin. As many as 17 percent of newcomers to Oberon had moved to be with their partners (Table 7.1) which also explains the more youthful migration into Oberon. Several couples had moved into Oberon to be closer to aging parents and grandparents. Michelle had moved back to be with her family after divorce from an abusive husband. Kinship was a powerful influence on migration and choice of destination, and kinship meant familiarity with the town. In 2006 fully three quarters of all newcomers had some family attachments (Brown 2006: 105-6). Moreover 'there is a bit of a cycle, where you leave Glen after school, go to uni[versity], work, then when you are at childbearing age all the negatives of Glen become positives. The small town is a bit of a backwater but it is safe for your kids' (quoted in Cuomo 2008: 68). Some migrants were simply coming home.

In the end for many the choice of Glen Innes or Oberon was somewhat arbitrary. Necessarily there was only limited time for a detailed analysis of particular places hence a degree of familiarity and friendliness went a long way. The internet was an important means of assessing facilities and services and real estate websites were invaluable for assessing property prices and availability. One couple found Glen Innes simply by searching real estate websites over a wide area of New England and continually being impressed by what was available around Glen Innes, prompting a visit to the area. A number of people had once come from those areas or had passed through the area at some time previously and been impressed.

While economic factors absolutely dominated migration, led by accessible employment and housing, they were strongly associated with intended lifestyle changes towards quieter less stressful lives, where kinship and community were more important. As perceptions of friendliness and community indicated, finding Oberon and Glen Innes was sometimes a matter of chance, though for many it followed some degree of familiarity and past associations. A wide range of people have moved into the two towns, cutting across demographic and class structures, employed or retired, from the state capitals and elsewhere. Oberon had rather more youthful newcomers drawn by jobs; Glen Innes had older newcomers anticipating retirement. In both places, some had taken years to decide to move; others had come almost on a whim. Some had been more or less forced out of previous places by rising house prices, others had the luxury of choosing acreage at will. Homes were bought, rented or inherited. Some people had been transferred, usually within the public service, and were not necessarily influenced by the place itself. Some were long familiar with Oberon or Glen Innes, a few having earlier chosen to move there and initially invest in land or property; others were swayed by chance meetings at events such as CW. Even the catalysts to migration varied. Distance played a part. A substantial number were returning to their roots, while a few creatively imagined country roots. What tied most together was some desire for a more relaxed lifestyle in places where they believed community to exist. Diversity otherwise prevailed.

New Lives?

Almost all those who had moved to Glen Innes and Oberon had come from larger centres and anticipated a different kind of life, even when they had been transferred there as part of their job. Since about a third of all the migrants had moved from places that were not much larger than Oberon and Glen Innes it might have been unrealistic to expect significant changes after migration. As much as anything else new residents anticipated a sense of community that they perceived as missing elsewhere, particularly in the larger cities that many had come from.

Some two thirds of those who moved into Glen Innes or Oberon, even those who had officially retired, were influenced in some way by employment issues. Many, especially in Oberon, had secured a job before migration, notably in the timber mill but also in the public service as teachers, police officers, wildlife rangers and ambulance officers. For some that was part of a career progression that brought higher wages and better housing, but perhaps not a commitment to Oberon. As one timber worker observed: 'The fact that I actually earn more money working here is an added bonus and helped in the decision-making process. We didn't make a conscious decision to move to Oberon other than to take advantage of a situation because we felt like a change' (quoted in Cuomo 2008: 75). Several newcomers worked in Bathurst or even Orange, because of the limited range of jobs in Oberon.

Glen Innes, without a single major industry, offered a more diffuse range of employment opportunities, and several newcomers had changed careers, some working in lower–paying jobs because of the lack of opportunities. Each of those who had taken lower paying jobs were undismayed, arguing that lifestyle gains compensated for reduced income. Three migrants to Glen Innes who had officially retired started their own businesses, perceiving this as a challenge, a means of remaining active and a service to the community. Others took up part-time work with the council or in the retail sector. As one newcomer stated: 'I started to work at the pharmacy a year ago even though we have been here for three years ... because I didn't really know anyone even though my husband is active in the community. Since working I have met a lot of people and made lots of friends. You can be a bit insular if you don't get out there and be involved' (quoted in Cuomo 2008: 75). Changing employment structures were thus at least as common as continuity, and for many were a major gateway into local life and community. By contrast reduced rural housing costs and thus mortgage payments meant both partners no longer necessarily had to work. Some women gave up full-time work, stayed at home or worked part-time. For some that meant extra time and an opportunity to consider extending families.

Most newcomers had moved into the towns themselves, though almost half (47 percent) of migrants to Oberon and 30 percent to Glen Innes lived outside town on rural properties. More affordable house prices enabled most people to become home owners; median house prices in Oberon and Glen Innes respectively were $190,000 and $160,000 in 2008 when they were $500,000 in Sydney. Within the towns, houses were usually single storey homes on classical Australian quarter acre blocks in tree-lined streets; invariably both houses and gardens were larger than in previous locations. Urban life otherwise gave newcomers familiar access to urban services such as water, electricity, sewerage, postal deliveries and garbage collection. Other services – such as shops and banks – and employment, were within a five minute drive or a 15-minute walk. Sports facilities were rarely further away. Many saw this as the best of both worlds: standard services but in more beautiful surrounds, and some experienced the pleasant novelty of walking to work, and having more time to meet neighbours.

While a couple of new households lived on substantial agricultural enterprises most of the 40 percent of households who had not moved into the towns lived on acreage, which usually meant ten to 50 acres, often not far from the urban fringes. For these households the change of life was often considerable. Much of that was experienced as greater freedom away from the confines of suburban life and the quarter acre block. Living on acreage gave children space and there was room not merely for a garden but to plant fruit trees or have a more demanding hobby farm. Some had significant aspirations: 'I want to put a dam down the bottom of my property later so I can do trout fishing whenever I want ... and no-one can tell me it's against the law because it will be my dam' (quoted in Cuomo 2008: 81). Such pleasures were not unalloyed. Many planted rows of trees as windbreaks and to hide their neighbours and other properties. Seclusion was what counted. Being on acreage meant no access to town water, garbage collection, gas or postal services. For those who valued seclusion that was no real disadvantage, yet others complained about the time-consuming and constant need to find firewood, the dangers of snakes, the annoyance of actually maintaining a large block and the frustration of constantly having to mow the grass. Even when they sought to do so the distance between properties meant that neighbourliness was more limited.

Newcomers' perceptions and anticipations of what it meant to be rural were embedded in most assumptions about migration. For many this was a key element in considering migration. Rural was often perceived simply in the physical landscape, especially the rural topography of wide open spaces extending to the horizon, and the sight of sheep and cattle: 'From my house I can see the rural outlook onto the commons and the hills with sheep in the background. I was looking for the view. I didn't want to feel locked in' (quoted in Cuomo 2008: 89). Newcomers had become more attentive to the weather and weather forecasts, since the impact of drought on the landscape was quite visible. Streetscapes too were different: 'It is much quieter, no street lights and no stress. When you walk down the street you are not pushed around' (ibid).

The physical ambience was accentuated by a somewhat different social world: 'You know it's a country town when you walk into the supermarket and see men with cowboy hats on and their utes out the front' (quoted in Cuomo 2008: 89). Utes (utility trucks), with cattle dogs in the back and 'cowboys' at the wheel, are prominent essentialist images of rural Australia. Various events such as intermittent farmers' markets, fairs and, in Glen Innes, the Celtic Festival (established as much as anything else to encourage tourism) enhanced the notion of difference: 'It's country because we have the Celtic Festival and the markets every month in the park in town where you can buy locally grown fruit and veggies', 'I like to go to the growers' market in town. I like to get the fresh peas that I have to shell myself; it is hard work but it is worth it because the peas have a beautiful flavour' (quoted in Cuomo 2008: 90). In such a small way, the 'hard work' of shelling peas became a metaphor for the simple sustaining pleasures of rural life.

The attractions of small scale can be considerable. One new resident liked the fact that Oberon had 'no traffic lights and only two pedestrian crossings'. The attitude and behaviour of established residents gave regional Australia a different ethos: 'I guess you see the country lifestyle in how people act. Like no-one is in a rush to be somewhere; they are not constantly looking at their watches' (quoted in Cuomo 2008: 90). Life was calmer and more relaxed. The friendliness and casual attitude of established residents were also perceived as markers of real country towns; in Oberon especially reduced noise, traffic congestion and crime gave many newcomers a feeling of safety and the confidence to leave their houses and cars unlocked. For newcomers from metropolitan areas pollution was much reduced and journeys to work minimised. Reduced commuting times made many feel healthier. As one newcomer pointed out:

> I was travelling four hours a day but now I can walk to work ... I don't spend money like I used to in Sydney. I was always popping out for coffee, juice or lunch ... even though I am earning less than I did in the city, over a third less ... it was worth it for what I gained mentally and physically (op cit: 76).

Time savings could be devoted to families and themselves. Even when migrants retained similar jobs, the working environment, and its social context, were more relaxed and they experienced less pressure, of deadlines, numbers (in the retail industry) and unrealistic expectations. Workplaces lacked some of the hierarchies and anonymity that made city employment more stressful; many welcomed the fact that their bosses knew who they were, which gave them a greater personal connection to the work, and the outcomes of that work were more visible.

Yet perceptions could change in quite a short time. Country towns that were perceived to be largely crime free turned out not to be so, and even limited urban growth had its detractors: 'When I first moved here [five years ago] the town was quiet and trouble-free. But over the years it has totally changed. Now there's too many people and more problems ... I don't like the town as much any more ... I need to find a quiet town like Glen Innes was before' (quoted in Cuomo 2008: 91). In contrast some of those who had once found country towns idyllic in their simplicity, and havens from urban pressures, found that life could be stultifying and boring, even a 'social desert' that displayed 'small town syndrome [with] discrimination in terms of race, age, sex and intellect'. Even patrons in pubs, supposed central places for conversation and community, could be stand-offish (Brown 2006: 106-7). Small towns were not necessarily melting pots.

Finding a Place

The majority of newcomers in both towns highlighted the role of community as a motivating factor in migration. Either they had experienced this in the context of making decisions, hence leading them to choose Glen Innes or Oberon, or they anticipated that it would be one of the beneficial outcomes of migration. Most wanted to live in a town with a strong sense of community which made then feel welcome: 'We were looking around at a few country towns ... When we came to Glen Innes we felt a different atmosphere. People were so friendly and really happy to help you. We wanted to be part of that' (quoted in Cuomo 2008: 69). Loose notions of friendliness and community had influenced choice of destination. First impressions counted a great deal: 'People here are friendly, like when we first came here, people would say hello and were friendly to you', and 'We were walking down the street and a kid said hi ... we thought that if this is how kids are then this is a good place' (quoted in Cuomo 2008: 68). Newcomers believed that this initial friendliness would be indicative of the openness of the community to new residents.

Many of those who had moved with the hope of finding connections with neighbours, perhaps as a prelude to connections with place, found neighbourliness more evident. A couple who had moved to Glen Innes observed:

> In Brisbane we were there for 16 years, we spoke to our neighbours now and again; you might say hello once in a while but nothing more than that. Our neighbours here have been very helpful. When we first came they brought us a bottle of wine ... we get along with them so well it's made us feel part of the place (quoted in Cuomo 2008: 7).

Such neighbourliness could be a matter of luck, and children often played a crucial role in establishing positive links. The idealised close-knit social relations of community could turn out to be insular, introspective social networks that could isolate newcomers. 'I think country areas are very friendly to a certain extent; there is an openness, friendliness and acceptance to new people but it only goes so far, then there is a deep seated reserve' (quoted in Cuomo 2008: 92). Even though some households felt they had been carefully assessed before the 'breakthrough', the small town suburbia was largely positive. Over time the sensation of being an outsider usually faded as newcomers became familiar with the towns and their eccentricities and made an effort to become involved by joining local organisations or working in local shops:

> It has taken three years, and only now are people starting to approach us but it has taken a long time because they [the established residents] want to make sure that you are going to stick around ... I think because we made the effort to join committees and volunteered, now people come to us. They [the established

residents] will be polite and pleasant but you have to be active in the community (quoted in Cuomo 2008: 92).

Where newcomers had kin, involvement was much easier. In Oberon, particularly, newcomers liked the fact 'that everyone knows everyone' to the extent that one got 'nervous when she hears sirens because more than likely she will know the person involved' (op cit: 88). New residents felt part of the community if people recognised them or engaged them in conversation in the street or in shops. Such simple pleasures made a great difference, and were invaluable sources of social capital.

Yet that was not always so. Several newcomers had experienced a degree of hostility from established residents who saw them as 'blow-ins' who had changed the ways of the town for the worse, and would never be considered 'locals'. Some recognised a logic in that: 'They think we come up here and want to do everything our way … I guess they blame us for the increase in property prices because we come up here and buy properties which means that their children have trouble buying a place' (quoted in Cuomo 2008: 82). But that sort of resentment was less evident in NSW than in the otherwise similarly changing small town of Castlemaine, Victoria, where previously established residents had moved away because of the shortage of cheap housing and rental accommodation that followed in-migration (Costello 2007, 2009). Resentment of newcomers was rather more obvious in rural areas where older values prevailed. Newcomers were sometimes accused of being unaware of the 'ways of the country': the etiquette and attitudes that exist in country areas, such as saluting others when driving, helping a neighbour when an animal had jumped a fence, or simply not being aloof from others when co-operation was expected. Some had built new off-the-plan houses, typically found in urban developments in the city, and demarcated their property boundaries through the construction of identical Colorbond fences. Such boundaries put them at odds aesthetically with the open barbed wire fences of the rural landscape that are part of 'community dialogue', where farm fences define property boundaries, but rely on individuals having a common understanding of the boundaries marked by particular fences, and that fences should be visually and physically permeable. Colorbond fences challenged such notions and formed an impermeable break between public and private spheres, that reflected newcomers' desire to belong to community yet establish privacy, practices at odds with the existing spatial and social norms of small rural communities, that emphasised newcomers' misunderstandings of the 'ways of the country' (Cuomo 2008: 100). Conflicts and tensions between newcomers and more established populations arose, as they did in the United States, from the 'changing identity of the community, conspicuous consumption of the countryside, increasing privatisation of resources, housing affordability, and issues of environmental conservation' (Ghose 2004: 529). The continued irony is that migrants tended to prefer a more idyllic and static countryside while established residents wanted some economic growth.

Recreation was part of migrants' new lives. Many felt not only that recreation was good for them and more time was available, but that by participating in formalised activities they could more effectively engage with local community and feel welcomed and accepted: 'I feel that I need to join these sporting groups … it is a good way to socialise' (quoted in Cuomo 2008: 86). Over 93 percent of newcomers to Oberon claimed to be involved in some social group or sporting activity. Sport was most popular, with football and rugby league leading the way for men. Tennis was more universally popular and golf had its devotees, notably with rather older migrants in Glen Innes, but some newcomers found golf particularly 'cliquey', with an inability to break into the more elite inner circuit. There was disappointment with the absence of gyms and activities such as yoga, and neither town had a swimming pool that could be used in winter. Some newcomers had attempted to establish more specialised recreational activities such as tai chi and belly dancing, but demand was too limited. In both towns renovating had become a new challenge, a form of recreation and a symbol of starting afresh. A handful of newcomers were drawn by rural pursuits such as hunting, fishing and mushrooming. Numerous small groups attracted some, from a Writers' Group, a Community Art Project through to the Patchwork Group. These latter activities tended to be for an older more sedentary population. In Glen Innes, Probus, an offshoot of Rotary created to offer retired business and professional workers opportunities to socialise, had become split into two groups; the original older group, mainly composed of 'locals', had over 100 members and a waiting list, hence a newer group had been formed mainly from newcomers. Social and recreational activities were not always quite as open to newcomers as had been anticipated.

Nor were social activities necessarily appealing to new residents. Although the *Oberon Review* recorded in 2006 that 'service clubs like Rotary, Apex, Lions and Masonic Lodge are always working on community projects … For those with individual interests, there is a community band, euchre club, garden club, needleworkers club, senior citizens club and weavers club' (quoted in Brown 2006: 64), few of these attracted urban migrants. A smaller number of newcomers, again mainly on acreage, had moved to avoid neighbours and secure a greater degree of privacy than was possible in denser urban areas. For them it was enough to know who their neighbours were but social relations were unnecessary. Not everyone wanted or valued local community.

Limited facilities were problematic for many, however much that might have been anticipated. One resident of Oberon disliked having to 'drive her son three times a week to Bathurst to play basketball'; another missed 'a movie theatre and a good quality restaurant'. Others recognised future problems, at least for their children, since the majority of work in Oberon is 'unskilled and low-paid like shearing sheep and mill work' (quoted in Brown 2006: 107). A similar conclusion was reached by a family in the small NSW South Coast town of Berry:

We realised that we actually liked art galleries and restaurants … We spent so much time driving our kids around that we thought it really limited their independence. In the city kids have more opportunities and can get around on their own … And you know what? I really hate gardening. I liked the idea of growing organic vegies but I hated doing it' (*Sun-Herald,* 19 April 2009).

They were returning to Sydney, as were a couple from the north coast:

We went because we wanted to experience country life: the freedom, smelling newly cut grass, having a vegie garden, riding horses. And while it was wonderful and a childhood they'll have in their bones for the rest of their lives, they were running wild a bit and it was time to inject more discipline and give them a more structured education back in Sydney (*Sydney Morning Herald*, 4 November 2007).

If neighbours were not always needed, some services were. Health services were limited, and many newcomers retained ties to specialists in their previous home areas, but access to general practice facilities was regarded as adequate, with reduced waiting times. Older people were more concerned about health facilities and some contemplated moving back to cities should their health deteriorate. Educational facilities were important for families; several households were unsure about the quality of local high schools but valued the primary schools, though many feared the time their children would move away for tertiary education. Transport facilities were limited and, despite local taxi services, for the few newcomers without cars this could be an irritation. Indeed Oberon's own skills audit, based on local people's perceptions, identified the lack of dentists as a major issue, followed by other health services, the weather, 'not enough police' and 'crime' (Oberon Council 2005).

Despite the significant reduction in housing costs the vast majority of newcomers felt that the cost of living was similar or even more than where they had previously lived. With relatively limited competition and considerable distances from the coast some goods are considerably more expensive in regional Australia. Several newcomers noted that it was much cheaper to shop in larger towns, such as Bathurst and Armidale, but travelling costs (and wasted time) cancelled the savings from such trips. At the same time many newcomers perceived that regional environments were less commercialised than cities and this was a positive benefit, since there were fewer goods, and less temptation and opportunity to spend money. In Oberon, the smaller town, there was disappointment over the limited range of fresh fruit and vegetables and the small range of shops. An irony of rural life can be reduced access to country produce, and even to markets. As many as 90 percent of all newcomers to Oberon travelled to Bathurst at least once a fortnight to shop, attend sporting events, music lessons or dine out. Oberon could simply be too small. In Glen Innes a third of newcomers travelled outside once a fortnight, a function of more local services and greater distances.

Happy Returns?

Few things can be more attractive than the notion of dissatisfied urban residents making a successful transition to rural life and, in the face of an alienating urban world, finding a new place in a smaller world of human proportions, dominated by a sense of community, seemingly absent in large cities. While smaller towns are said to be more attractive to older people seeking a slower pace of life, and a greater sense of community, more limited services can be a critical deterrent. But many former residents had, perhaps as part of a life cycle, returned. Other new residents may never have lived in either town, like Lauren (above), but had lived in similar places and valued the broad ethos of regional life. Others with no historical connection to either area waxed eloquent about the positive connection they felt with rural life: 'we just knew', 'I felt like I was home' and even 'I felt there was something drawing me here, I had dreamt of returning to the place of my female ancestors'. Yet another couple felt 'as if we are walking in our ancestors' footsteps' (quoted in Cuomo 2008: 63, 67) on the basis of the Celtic history of the region. Some had a strong sense that the country was where they belonged: 'I grew up in the country and I have always liked the atmosphere, the connection with nature and simple ways of doing things' (quoted in Cuomo 2008: 70). Kinship played an important role. Some newcomers were elderly retired people who had relatives in the towns and wished to be close to them in later life; likey Castlemaine (Victoria) where many moving from Melbourne already had some connection, usually family ties, with the town (Costello 2009), this ensured that adjustment was less difficult than it might otherwise have been.

Belonging?

Perhaps the greatest challenge for many is fitting into a new social context, where established residents are not always welcoming to new arrivals. After all many differences seem universal. In England there were broad tensions between newcomers in English villages over what might have seemed petty divisions, such as whether villages should have street lights or pavements, as outsiders simultaneously sought both more Arcadian landscapes and urban amenities (Pahl 1965, Connell 1978). In Navarre (Spain) the divisions between locals and newcomers created cultural dislocations that undermined 'the idealised communitarian experience and emotional investment made' by the newcomers (Escribano 2007: 39). In both contexts divisions became more evident when migrants became politically engaged. In England and in New Zealand new residents were prone to proclaim themselves as guardians of place and, at least in New Zealand, tended to 'articulate their position as "locals" more strongly that those born and bred in the place' (Fountain and Hall 2002: 164), who unsurprisingly resented such presumptions. Somewhat similarly in England it was the newer, middle-class migrants into English villages who 'were always trying to organise events in which the village could be seen as a public entity' (Strathern 1984: 103). In Spain active involvement in local politics was 'mostly considered as open confrontation ... an interference in the life of the village' (Escribano 2007: 39). In NSW, even when rural towns were demonstrably dying, as at Nimbin in the 1980s and Cumnock in the 2000s, outsiders were nevertheless unwelcome to some, symbols of a distant and alien world, disturbing local certainties and continuities. There are acute parallels with the lack of support in many small towns for festivals that seemed to smack of urban follies or sectoral interests, pointing to existing and entrenched divisions in regional Australia (Davies 2011, Gibson and Connell 2011), distaste for novelty and innovation, and no little conservatism and inertia.

In Oberon established residents argued that, after the lack of certain services, the worst thing about living in Oberon was that it was starting to change. An older resident observed: 'I love it in Oberon but it's changed a lot – lots of new home construction and city people moving in' while another observed the 'friction between newcomers and real locals' (quoted in Brown 2006: 64). Friendliness to tourists, who are consumers, may dissipate for migrants who may be competitors. As Baldacchino observed in Prince Edward Island (Canada), local people are less friendly when visitors become residents; the unfamiliarity of newcomers with the 'intricate and ascriptive social network and *gemeinschaft* ... can create serious difficulties of integration for those who are "not from here" [which leads to] exasperated newcomers, sooner or later, packing their bags and moving on, and thus reinforcing the uniformity of the local cultural space' (2008: 12). Once again it is not surprising that those who are most likely to succeed in regional contexts are those who have been there before, may well have connections, and also perhaps a sense of obligation and responsibility (Ni Laoire 2007). Though the rigid social divisions in small towns such as Barcaldine and Bradstow in the 1960s and 1970s (Montague 1980, Wild 1974), have long gone it remains a familiar story that migrants have to be in regional areas for long time periods before they can conceivably be seen as 'locals'. Almost half of tree-changers in western NSW and north-east Victoria 'felt that they did not fit into the community, and that there was a sense that they had to have several generations born and buried in the area to be a local' (Angela Ragusa, quoted in Munro 2009: 3). Such 'requirements' are both insuperable and objectionable to those who would seek to 'fit in'. Belonging suggested stability, permanence and a suspicion of change; newcomers potentially challenged such assumptions. Yet newcomers 'didn't know how to connect with locals and some didn't try, they didn't know where to start ... Many would like to move back to the city, but can no longer afford it and become quite disgruntled about that' (ibid). For a variety of reasons not all those who have moved to country areas choose to stay. No doubt some of the quickly dissatisfied had already moved on or back.

No migration to a new town is likely to be without some challenges and even disappointments. Equally, unless newcomers have made quite foolish decisions, it should provide pleasant experiences. Relocation is rarely a hurried decision, especially for those who are relatively well-off or contemplating retirement, but may extend over decades, until the right alignment of circumstances, so that destinations are scarcely unknown. Despite probably inevitable teething problems, the majority of newcomers, 63 percent in Oberon and 70 percent in Glen Innes, were well satisfied with their move. Not one person in Oberon aged over 40 had anything negative or even merely neutral to say about the town, and in the 'unlikely' event of their moving again it would be to another country town (Brown 2006: 108). Many felt it had exceeded their expectations, and commented that urban friends were envious of them. Even higher proportions of the newcomers intended to stay. Personal attitudes and personality played a part; some form of joining in was particularly important. Less than 10 percent were dissatisfied, with younger people who had moved primarily for employment reasons least content, because of the lack of social activities, entertainment and the conservative nature of small town life. Almost all of those who intended to leave were households who had moved for employment and saw their time in these parts of regional Australia as part of a process of economic and social mobility through the workforce. Small towns cannot possibly satisfy everyone equally and are stepping stones for many. Indeed at the 2008 Queensland Expo one small town stressed: 'We want people to come even if they only spend part of their lives here'.

While lifestyle persists as an explanation for migration, and lifestyles certainly change after migration, the evidence from regional Australia indicates that lifestyle is subsidiary to employment and housing, and thus to financial issues. Outside coastal retirement areas, such conclusions are widespread, as in the United Kingdom and the Netherlands (Bolton and Chalkley 1990, Halfacree 2004, Vergunst 2009). Yet even in two small towns of similar size there was a wide spectrum of migrants and outcomes of migration, with migration from a range of locations, for a variety of reasons and into diverse new lifestyles. For some, migration took them to suburbs not unlike those they had moved from, basically non-metropolitan forms of urbanisation (Burnley and Murphy 2002), while others 'went bush'. The resultant lifestyles varied, with positive phenomena (time and relaxation) and negative (reduced services), rated in different ways. Lifestyle may not have been the key motivation for migration, but all migrants expected some positive lifestyle benefits; the countryside was not expected to be penance for mobility. And nor was it. The many variants of tree change migration evident in Glen Innes and Oberon brought positive gains for newcomers even where they were little more than a 'change of pace' and a sense of community. Migration only exceptionally heralded radically new lifestyles.

While migrants to Glen Innes and Oberon faced certain challenges to becoming established – perhaps because they were unlike established residents, or they expected too much of small places – some new migrants experienced enormous success. Such residents have been signalled out in the pages of the local newspapers as 'success stories' that both demonstrate the potential ease of transition for some and the kind of new residents who are successful. These stories are examined in the next chapter.

Chapter 8
The Good Resident

Most towns attending the Country Week (CW) Expos brought along special editions of their local newspapers, which played a particular role in highlighting the positive features of regional life, often focusing on the successful transition to the town of some of its recent residents. Some towns, like Narromine, Glen Innes and Rockhampton, even brought out special booklets centred on the experiences of newcomers. Together they constitute a series of controlled discourses on the nature of migration and country residence, that stressed the positive attributes of the new migrants and country life. Other publications, such as *Live the Dream*, and CW's own promotional literature and website, similarly emphasised the successful outcomes of migration, and highlighted those deemed successful. What that success amounts to and who have been able to achieve it can now be examined, through an analysis of almost a hundred such stories from both states (though most are from NSW where the longer history of Expos yielded more stories) and from various local publications. None of the new residents had been in regional Australia for more than five years.

Most stories were several hundred words long, and even longer in Rockhampton's 48-page 'Central Queens*Land* of Opportunity. 12 Lifestyle Adventurers Living The Dream' (Rockhampton Regional Development Limited 2007). Most therefore, like the extended case studies from the CW website (below), offered multiple reasons for migration and for successful adaptation, that often closely paralleled CW themes. Other than the seminal stories from CW's own website they cannot be repeated in any detail but their key features are distinguished and analysed below.

Telling Stories

The most comprehensive accounts of new migrants to the country are those from CW's website (July 2008), which well demonstrate the key themes of success and emphasise the positive characteristics of the newcomers. They are thus worth drawing on in detail as valuable extended case studies of what constitutes 'success'; they replicate the themes that permeate CW's promotional literature, and are emblematic of its perception of success. Accompanying photographs portray and emphasise crucial themes, as on the cover of various Expo brochures, dominated by young families and couples, happily relaxing and living well in verdant countryside (Figure 8.1) effectively summarising both the target audience and the ideal migrant groups, within an idealised landscape. Single people, and even older people, were neither targeted nor so lyrically portrayed.

Figure 8.1   Country Week Expo brochures

*Source*s (clockwise from top left): 'Live the Dream', July 2007 Queensland brochure cover; 'Make the Move' 2008, NSW brochure cover; 'Live the Dream' August 2007 NSW brochure cover; 'Country and Regional Living Expo' brochure cover, NSW 2009. All images used courtesy of Country Week Pty Ltd.

In contrast to many stories of migrant households, accounts of the successful movement of large businesses are extremely rare (compared with many accounts of small businesses such as cafés). One of Peter Bailey's strongly held personal views was that greater success came through the migration of significant economic enterprises, rather than simply of households, because of their multiplier effect. In 2008 one rare success story was highlighted: the movement of a fireworks company from the fringes of Sydney to the township of Marulan

> Foti Fireworks is the largest and most awarded fireworks manufacturer and display operator in Australia. The family company is located in Leppington, not far from Liverpool in Sydney, but the continued residential growth in the area has made it impossible to continue operating in a metropolitan area. 'Our operation needs buffer zones, large areas to operate successfully. We're being encroached upon by suburbia', says Company Director Vince Foti. The Foti Fireworks Company decided to look further afield and finally targeted a site at Marulan near Goulburn. Vince Foti says 'Marulan has given the company all the space it needs'. Mr Foti says the company will be taking key staff with them to Marulan and will be asking all their staff to move with them. He said 'We will definitely need locals at Marulan to come and join us. We see it as a win win situation. We are able to move our factory to an area that is much more suitable for our line of work and at the same time we are able to offer employment to the people living in the area'. Vince Foti says the company picked Marulan because it's close to good transport and only an hour and 15 minutes from the present site at Leppington. Mr Foti is excited about the company's move to the country. 'It's great, as a business moving to an area like this you get to grow with the community'.

While this story is more about an anticipated positive outcome, and Foti is unusual in requiring considerable open space, it nonetheless emphasised several themes that recur with household migration: family, community, transport access and proximity to the point of origin. Almost without exception, all other stories, whether of CW or in the local media, focused on families who had successfully relocated. Some, like the Burkes, had developed an earlier small business in a new area, but most of those who appear in stories worked for others. Again it is useful to start from CW's 2008 website with its four detailed case studies in their entirety.[1]

Tatyana and Dennis Yule – Tree Change Story
Tatyana and Dennis Yule moved from Sydney to Armidale on the state's Northern Tablelands 18 months ago with their two sons Frazer, 11 and Connor, 10. As Tatyana writes, this major move has opened up a whole new world of opportunity for the family. 'We were a couple significantly entrenched in Sydney, feeling life

---

1   Website extracts used courtesy of Country Week Pty Ltd.

was content and satisfactory. Good jobs, in the middle income bracket, 2 kids; living in an acceptable suburb in the inner west of Sydney. We were making good money, provided there was some overtime every week, but could not afford to buy our own house in Sydney unless we were willing to commute 2 hours each way every day or not be choosy about where we sent our kids to school. We picked the suburb in Sydney where we lived for its school so that we lived 'within the zone' ensuring our children's enrolment. Average house price in our suburb was $650,000 to $800,000. Average rent for a very basic 3 bedroom, $375 to $500 pw with no garage. With the cost of transport, tolls, parking, petrol, childcare, insurances, rent and the other basics, we felt like we were going around in circles. Moving to Armidale was a risky and intriguing proposition for us as neither of us had ever lived in a community under 700,000 and had no family or support network here. Job prospects, infrastructure, schools and education for the kids were top of our priority list. We were seeking a better quality of life, one in which we could enjoy our family and not be working endless hours or stuck in traffic for endless hours. One where when we actually reached the comfort of our own home we didn't feel exhausted. Since moving to Armidale 18 months ago – I secured employment within 3 months, albeit on less money. We bought our own home and are paying $50 less on our mortgage than our rent was in Sydney. Our costs for insurances, childcare and transport have been halved and we are renovating our home. The most precious gift to us is the 3 extra hours on any given work day which are now ours which were once spent lost in 'the twilight zone of commuting to and from ...' and can actually do things on a weekend as a family. The school our children attend is well resourced and committed to an equally high academic standard as their previous Sydney school; the parents have a sense of community and contribution. We still get our "city fix" every 3-4 months for a weekend in Sydney or Brisbane or with friends and family, the "bright lights" and shopping of course. For the first time in years, we are not just walking a treadmill of basic economic survival and can see possibilities. My husband has been working in Sydney on occasion and says he loves getting back to Armidale, the peace, the beauty of the area and our home. 18 months after our move, our children are happy and busy and we are more settled in the community'.

Check It Out – It's Worth It

When Merrilyn and Maurie Jackson decided to move out of Sydney to somewhere in the country they made all the right moves. First of all they attended the Country Week Expo at the Rosehill Gardens Events Centre last year, and once there, went about their business in a very methodical fashion. They went and visited all the stands at the Expo meeting and talking with people from country and regional centres all over NSW. They collected sample bags at all of those stands, and where they met people they liked, they put a big tick on the bag. When they arrived home they took out a map of NSW, and measured out a radius that was roughly three and a half hours travel time from their home at Plumpton near

Blacktown. When they compared the ticked bags to the map, they found they had a lot of places to choose from. They visited about half a dozen towns before an invite came to be the guests of the Cootamundra Shire Council at the town's annual Wattle Time festival. 'Having lived in the city all our lives and being a little sceptical of anything for nothing, we thought all the Cootamundra people will be on their best behaviour for visitors during their festival, so we went and checked the town out the week before', says Merrilyn with a laugh. Everything they saw that day they liked and the Jacksons were back the next week for the festival. 'We were talking to a lovely old lady in the street at Cootamundra during the festival and when we went to say goodbye, she actually gave me a kiss and a cuddle. Straight away I knew there was something special about the place,' Merrilyn remembers. Finding their dream home on the outskirts of Cootamundra was also like it was meant to happen. The couple lived in a caravan park when they first moved to Cootamundra but strolling up the main street one evening, they saw a house in a Real Estate agents window. They called the agent up, and even though the next day was Father's Day, he left his family to show the Jacksons over the four bedroom home on five acres that had caught they eye. 'That wouldn't happen in Sydney', says Merrilyn. 'The people in Cootamundra are so friendly; the whole atmosphere in the town is wonderful, it's just right!' It didn't take long for Merrilyn and Maurie to make up their minds and now that house on five acres on the outskirts of Cootamundra is home. 'Compared to where we used to live in Plumpton, the crime rate in this area hardly exists. In Sydney, there is lots of vandalism, graffiti is everywhere and there is never any shortage of hoons. There's nothing like that at Cootamundra. We feel like we're in the middle of nowhere with no neighbours in sight.' 'The grand children love it here, our health has improved and I'm even turning into a housewife and a cook', giggles Merrilyn. 'These days I find I'm making homemade soup and biscuits. There just seems there is always something to do'.

It's The Best Thing We Have Ever Done

That's the view of Mark and Anne O'Connor who moved from Sydney to Tamworth just over three years ago. Mark, an accountant, and Anne a teacher were living in Thornleigh in Sydney when they decided to make the move. Their first two children had been born and the pressures of living and working in the city were starting to take their toll. 'I was working long hours and at the end of it, I sat in the car for 50 minutes before I was able to reach home and be with my family. Even now, our two eldest boys don't go to bed until nine o'clock because they just became used to waiting up that late so I could say goodnight'. says Mark. Mark and Anne have three children; Jamie 7 and Nick 4, both born in Sydney and Ben, who is almost two and who was born in Tamworth. In fact little Ben might not even have been here except that his parents decided to move out of the city. As Anne says, 'If we were still living in Sydney, I doubt we would have had a third child. We would have struggled and I would have needed to go back to work.' Anne has no doubt that their decision to leave Sydney was the right one. 'The move has brought us closer together as a family, Mark and I are closer and Mark is closer to the boys than if we had continued to live in the city'. The O'Connor's were able to sell up in Sydney and buy a home on three acres in Tamworth. After ten years of equity in their Thornleigh home the changeover was more than favourable financially. While the move has brought great personal assets into the O'Connor's life, it has also been extremely successful professionally. Mark is a Director of Warburtons Charted Accountants in Tamworth, one of the city's most successful firms. 'The opportunities for professional people are extraordinary', he says. 'In addition to our local clients, we are attracting more and more clients from metropolitan areas. Because our overheads are much less than our city counterparts, we can charge half their fees but with technology and on-going training and our experience, we offer the same level of expertise and professionalism.' Mark thinks city people have an unrealistic view of what country areas have to offer. He says, 'City people need to take the trouble to have a good look at what is on offer in country and regional areas. They will get a huge surprise'. 'There is great opportunity for professional people to find interesting and varied work in country areas. At our firm alone, we could put on two additional qualified people tomorrow.' One of the great bonuses for Mark and Anne is that they were both Tamworth people originally and so returning to Tamworth after a 16 year absence has put their children into direct and regular contact with their grandparents. It has allowed them to tap into a family support network that was missing in Sydney. Mark often goes home for lunch during the working day, something he would not have even thought about in Sydney. He finds the time to attend school functions with his children, again something that would not have happened in Sydney. For Mark and Anne the work/life balance is back on an even keel. There's not doubt in the minds of Mark and Anne O'Connor … moving from Sydney to Tamworth was just the best thing they have done.

The Burke Family

A move to the northern NSW city of Armidale two and a half years ago has opened some amazing doors for the Burke family. Gordon and Dominique Burke and their sons Nicholas and Alexander left Sydney's Barrenjoey peninsula for an eighteen hectare property near Armidale and a whole new way of life. Gordon and Dominique had owned six cafes in Sydney. As Gordon says the company colours were red and white to denote the couple's passion for good coffee. Since coming to Armidale the Burkes have a new passion and this time their company colours are in the cool blues and silvers of the information age. Gordon and Dominique found a need and are filling it. For years Gordon travelled all over NSW but failed to find accommodation that he felt was suitable. So there was only one answer. Create something that filled Gordon's own expectations. The couple have established Executive Oasis – accommodation for travelling executives that is more a home than a motel room. What's more they are on the cutting edge of technology with all their transactions with customers being on the internet.

Customers log on to the Executive Oasis site and choose the location of their accommodation. The financial transactions are completed electronically and the guest is given a pin number that takes the place of a key. All the guest has to do is arrive at the accommodation, punch in his supplied pin number on a key pad by the door and gain entry to his luxury home away from home. As Gordon says, 'Our motto is comfort, quality and consistency. We are providing high class, smart accommodation which even includes high speed unlimited broadband all included in the cost'. 'There are enormous business opportunities in NSW. I am living proof of that' he says. Gordon, Dominique and the boys moved out of Sydney to be closer to Gordon's parents who are living in Armidale. The couple are always looking for opportunities. Gordon took a job with Telstra which allowed him time to see what was available in the area. Gordon says 'There is a wave of migration happening out of Sydney to regional areas. These are people who are well paid professionals looking to improve their lifestyles.' Gordon and Dominique are extremely pro Armidale. 'We love the people in Armidale. In fact I did some of my schooling here and still have friends from all those years ago', says Gordon. He believes there is a special quality of friendship, a real 'genuineness' about country people. 'When we lived in Sydney people rated us by the car we drove, by the school our children went to, by the suburb in which we lived. In Armidale people care about who you are, not what you've got. There's a real generosity in the people'. As Gordon says in Armidale you have all the benefits of Sydney without having to put up with the crap! 'There is a real feeling of community here, that's what it's all about.'

Collectively the many recurrent CW themes are here: 'urban anxieties' (such as security and congestion) and the much vaunted contrasting country 'opportunities' (time, family and space, employment, education services, community and friendliness) with savings from cheaper house prices, enabling greater peace of mind, bonuses – better work-life balance and improved health – alongside continued access to the city. Three of the households had at least two children, while the Jacksons, in the much smaller town of Cootamundra, appear to have retired. The others were in the large regional centres of Tamworth and Armidale. Only the Yules even hinted that they might have had initial problems, in the lack of kin and a support network, but such problems were quickly vanquished. Just one household had any previous experience of country residence.

The local newspapers in their special CW editions not so much take these themes further, often using CW slogans as headlines, but reiterate them. Here too however most of the 'good residents' have taken up, and responded to, themes that are remarkably similar to the advertising and marketing. Thus one family moving from Robertson (Kangaroo Valley), an unusual move in being from another rural area, were said to have found that:

The highlights were – no traffic lights, three country style pubs, a couple of clubs, a Chinese restaurant, a classic Greek café serving hearty delicious hamburgers and fish and chips, an ambulance station, a hospital, a police station, a huge blue cathedral-like Catholic Church and various other places of worship, two primary schools and a High School – to name just a few things in beautiful, historic Crookwell (*Crookwell Gazette*, 7 August 2008)

Such an unsubtle gazette of improbable highlights was otherwise rare. More realistic was the story of one family who moved from southwestern Sydney to Grenfell and whose changed life was reported under the headline 'Meet the Edwards Family. They made the break and are now living the life the city-bound people can only dream of'. At the core of a longer story:

However the real joys of making a tree change is the lifestyle overall. The clean air, the casual pace of life, the great award-winning schools from pre-school through to high school with their extensive grounds, no traffic jams or five hour commutes every day (there are no traffic lights in Grenfell), little or no crime, a safe environment for kids to grow up in and relatively cheap housing and fresh food made the choice ... a no-brainer (*Grenfell Record*, 1 August 2009).

Many regional newspapers and other publications summarised various versions of the tree change in similar ways, and in much greater detail, hence these stories have been disaggregated into the key themes.

A Sense of Place and Scale

Small is beautiful might have been the theme of many stories, and small size spilled over into many other benefits: community, friendliness, time savings and easy access to facilities and rural scenery. 'Space, air and being at one with your environment is great' (South Burnett Regional Council 2008). In Armidale: 'The major benefit of living here was the open space and to be among the wildlife. Having a pet in Sydney was not an option, but now I have a large backyard and a newly adopted Maltese X Poodle named Kevin' (*New England Focus*, August 2009). In the smallest places it was possible to get to know the whole community, be known by them and have personal relationships with shopkeepers.

Journeys to work were a pleasure rather than an endurance test. Both the Yules and O'Connors saved vast amounts of time for more worthwhile activities, not least recreation and family time. The Pardys, who together travelled a daily four and a half hours to work within Sydney, both now walked to work in Cootamundra (*Cootamundra Herald*, 1 September 2006). 'In Armidale it takes me for or five minutes to get to work. People talk about the cost of petrol, but everything's so close. Not having the hassles of commuting you find in the city frees up a lot of your time. I was able to take up recreational pursuits when I came here' (*Armidale Independent*, 6 August 2008). A particular theme in time savings was that sport and recreation were now possible and the work-life balance was forever altered. That enabled new or revived pursuits such as golf and tennis or simply, like Merrilyn Jackson, the time to cook and bake. A new sense of place could be translated into reduced pace and more relaxation.

Community and Friendliness

Warmth and friendliness were obvious to some before they had even properly considered a move. Merrilyn Jackson was literally embraced at the Wattle Time Festival in Cootamundra. A household looking at the possibility of moving to Oberon 'talked to every person in every shop, and what struck us was how friendly everyone was – there wasn't anyone not prepared to talk' (*Oberon Review*, 4 August 2005). A Sydney couple on their first day in Cootamundra found that it 'was perfect, we loved the look and the feel of the town immediately' (*Cootamundra Herald*, 1 September 2006). Another in Grenfell: 'the moment we drove into the town we got a warm feeling and instantly loved it' (*Grenfell Record*, 9 August 2009). In Parkes 'I can honestly say we liked what we saw right from the outset. Everyone we spoke to was really friendly and people bent over backwards to help and make us feel welcome ... We might have been here only eight months but already we feel part of the town' (*Parkes Champion-Post*, 7 August 2009). First impressions were thus later consolidated.

Friendship was not always perhaps quite so immediate though it was assured. After moving in the Burkes found 'genuine friendship' from 'people who care about who you are' rather than the materialistic, superficialities of Sydney. In Yass 'The people are just terrific. Yass has a lot to offer, everyone is so willing to help each other and nothing's ever too much' (*Yass Tribune*, 6 August 2009). 'Everyone's very friendly. They wave at you in the street' (*Boorowa News*, 6 August 2009). 'It's really nice when you go to the golf club, or to the footy, that you know people and you aren't just another person in the crowd' (*Narromine Shire Council* 2007). In Hughenden 'neighbourhood watch just come naturally ... people help you out for no reason at all ... you'll get a feeling of belonging because people are open and friendly' (*Hughenden News*, 19 September 2008). Initial warmth and friendliness blossomed into more substantial relationships. For most new residents, as in Grenfell, 'their social life is more fulfilling than it ever was in Sydney and they believe these are friends for life. That happens in a town where everyone knows everybody and they rely on you to be part of the fabric of daily life' (*Grenfell Record*, 1 August 2009). Fitting in came with seemingly little effort. Only in Narromine was it even suggested that 'people who want to move to the country have to be prepared to be involved' (*Narromine Shire Council* 2007). New residents were usually entirely willing to be involved.

Newcomers usually joined in some formal social activities, often sport, since there was more time: 'on a weekend you will see Daniel playing second grade with the Boars rugby union team, or out on the golf course hacking around' (*Parkes Champion-Post*, 7 August 2009). Church groups were sometimes hinted at, but were only mentioned indirectly, and politics was entirely absent. Organisations like Rotary tended to exert a particular attraction to retirees, who found a broad social life: 'I am busier here in retirement than I ever was in Canberra; it's very satisfying to be involved in a small country community' (*Boorowa News*, 6 August 2009). Nowhere lacked opportunities to enjoy being involved.

A Place for Children ...

Children benefited from the country for many of the same reasons as parents: more time and space, fresh air, recreation facilities, less threat of violence and therefore greater security, alongside good education, of standards equivalent (and perhaps superior, through small classes) to that of the city. Children were safe. 'We have lived in over 30 places in six countries in 40 years and this town is by far the friendliest and safest for our kids ... the most supportive of them all with a close community feel and everyone making us feel included. ... Cunnamulla is a very safe place to bring up children – and there is a sense of trust in the community' (Cunnamulla Queensland, Paroo Shire Council 2007, ms). 'Oberon is a nice little community, especially with children, and we don't have to worry about them walking down the street. Everyone knows everyone so they look out for each other' (*Oberon Review*, 4 August 2005). 'Grenfell is safe, friendly and a healthier environment for our kids' (*Grenfell Record*, 9 August 2009).

Children were able to do the things that children were supposed to do. 'Their children have also settled in to the vigorous sporting life of the town with them playing soccer and Courtney has a horse on the 20 acres at home and is now a member of the Pony Club' (*Grenfell Record*, 1 August 2009). 'Kids around here can still ride their bikes to the park or the swimming pool and play, they can fossick around down by the creek, they can while away time developing their imaginations without adults constantly watching and worrying' (*Yass Tribune*, 6 August 2008). 'In Sydney there is no space for the boys to be boys. We are looking for acreage for the boys to run around and give them room to move' (*Armidale Express*, 16 August 2006). Only once were children themselves quoted, and they had the same perspectives: 'I like pre-school. The parks, watching football on Friday night ... being with my friends ... playing in the mud ... rounding up sheep on a bike' (*Yass Tribune*, 6 August 2009).

... and Families

The majority of stories relate to nuclear families with a couple of children; almost all the other stories related to families either with more children, or with fewer children but who had moved to the country to take advantage of more space and reduced costs to expand their family. Very few stories related to retired couples without children, and then usually in publications that featured several case studies, enabling an indication of the diversity of new residents. Only exceptionally, for example in the city of Rockhampton, where 12 case studies were printed together, was there any significant deviation from this family theme. Not only were families much the most important migrants but migration meant more family time, bringing parents and children closer together, as in the case of the O'Connors, and enabling weekends to become family time, as for the Yules.

Many families moved specifically to have more family time. 'Rebecca and I discussed a lifestyle move to the country after realising we were experiencing little family time together because of work commitments' (*Parkes Champion-Post*, 7 August 2009). 'The children have both settled in well at school and the best part is I'm getting to watch them grow up and be a hands-on dad' (*Parkes Champion-Post*, 7 August 2009). Family time could sometimes be enhanced by one parent stopping working because of reduced costs. Family life could also be strengthened in other ways: 'I now live five minutes from work, go home for lunch and am home with the family by 5pm' (*Cowra Guardian*, 5 August 2009).

Narromine, in their booklet of five tree changers, was one of only two places to feature a single male migrant, under the title 'Flying Solo', a 29-year old teacher posted from Sydney to the local high school.

> I thought I'd make sacrifices moving from Sydney to the country but I haven't. In many ways it is just like Newtown [a trendy, cosmopolitan inner city suburb] only there is one of everything ... I thought I'd miss the social scene in Sydney but I don't. The community is so welcoming and I've been able to meet a full spectrum of wonderful people here (Narromine Shire Council 2007).

While he had been posted to Narromine, the only other single migrant featured had gone to Armidale 'by mistake', having been forced out of the Sydney property market and chosen to invest 'temporarily' in Armidale, where his parents had previously moved. Three years later he had become established: though he visited Sydney once a month 'I don't know if I will ever go back now' (*New England Focus*, August 2009).

In 2009 the CW issue of the monthly *New England Focus* reported at length on a pair of business men with long connections to the region who had made a tree change from Sydney, in search of 'peace and quiet, a healthier way of living, the connection with nature and a close community of like minded people', in a place with a cultural life, good transport connections to Sydney and to the coast. While the story covered where and when they met, and how they had come to buy properties in Armidale and elsewhere, no mention was made of sexuality (and nor was there need to do so) but the implications were reasonably evident (*New England Focus*, August 2009). Perhaps only in a relatively large university town would such a story have appeared.

This lone story emphasised that implicitly country towns were not for minorities, whether gay, of 'ethnic' origin (including Aboriginals, though few were likely to have been tree changers) or single parents. While such minorities have moved to many regional centres, they have met challenges in some places, ranging from opposition to gays in Daylesford (Gorman-Murray et al. 2008) and more generally (Roberts 1995), to perceived 'hippies' in Nimbin and Denmark (Curry et al. 2001: 117) and ethnic populations in centres such as Griffith. Other than in Rockhampton, a city of 60,000 people, almost everyone had Anglo-Saxon names. Otherwise a Japanese wife and Estonian, Belgian, Indian and Italian surnames, were depicted in stories that, without exception, indicated their previous long familiarity with country Australia. In Rockhampton the expanding economy and city had drawn migrants from a number of countries (although almost had previously lived in a larger Australian city), including Sri Lanka, Iran (via Laos), Vietnam, Russia and China.

A significant number moved because of previous experiences or kin in the country, ensuring both that they were familiar with their new destinations and that, often, there were kin connections to ease the move. 'We have holidayed in New England for many years, with my parents owning land near Walcha' (*Armidale Independent*, 6 August 2008). Returning to roots and family was an intermittent refrain.

Services

Almost all the stories emphasised that services were readily available in the country and that only inessential, unnecessary services were absent. Since most stories were about young families the main focus was on the quality of education provision at primary and secondary level, and the more personal care in smaller classes. At Black Springs, outside Oberon, 'the school gives children opportunities and special attention that they may not receive in larger schools and the close-knit group of students means children play and learn together. It teaches people from a young age to associate with people from different age groups' (*Oberon Review*, 6 August 2009). 'It's appealing to have our children in a smaller school' (*Oberon Review*, 10 August 2006). 'The schools here are well maintained and the classes are a good size which means your kids get the attention they need' (*Grenfell Record*, 9 August 2009). Only in the exceptional case of Armidale, with its university, were tertiary facilities mentioned, and then mainly in the guise of the broader cultural implications for the town.

Only in the context of retirees were medical facilities given any real significance. Narromine is 'centrally located, only half an hour from Dubbo, and all the major services are there [including] a new medical centre with modern facilities and there isn't much you can't get locally ... especially those that aging people may need in the future ... A town like Narromine is appealing to retirees because it offers a quiet lifestyle with all the facilities and conveniences you need without having to live in a big city' (*Narromine Shire Council* 2007). There were perhaps implicit qualifications to service provision:

> If you come to Rockhampton expecting a big-city lifestyle only on a smaller scale, you may be disappointed. On the other hand, if you move from the suburbs, the differences aren't obvious. There are two McDonalds, two KFCs, three Red Roosters ... How many suburbs do you have to drive through in Brisbane and Sydney to get the equivalent? Really it's just that at the edge of the bitumen we have trees and paddocks, not more people and buildings (Rockhampton RDL 2007: 26-7).

Basic facilities were universally available, but subtle tastes were not always catered for, and in smaller towns, such as Oberon (but not Glen Innes), McDonalds too were absent. It was constantly reiterated that all the basic services were available, and any unusual items might be readily obtained from larger towns nearby.

Economic Growth

Small towns were not economic backwaters. The Burkes found 'enormous business opportunities'. Many others were able to establish businesses such as Electrical Services and Scootamundra (a scooter store) while their partners too found congenial employment, a concern of many at the Expos. 'Setting up our café in Narrabri was very easy and since opening our patronage has been fantastic. Business is steadily growing and I can only see this improving with the coal mining around the town and diversity in the local agricultural industry' (*Narrabri Shire Business and Residents Guide*, ms 2008). For one couple: 'we mainly aimed at Oberon because we thought it had a lot of potential for growth, and since we've been here this has been confirmed' as their business was performing better than they had anticipated and they felt this would continue (*Oberon Review*, 4 August 2005). A new resident in Armidale, who had taken over tourism operations for the locally based Fleet Helicopters, noted 'Armidale has a lot of potential to grow, through tourism and other business opportunities ... there's definitely the capacity for growth' (*Armidale Independent*, 6 August 2008). Successful business developers were enthused over the continued potential for expansion.

One business owner argued that business development was actually easier away from the state capitals: 'business development would be more difficult to achieve in a competitive market like Sydney' (*Muswellbrook Chronicle*, 8 August 2008). At Warwick a former builder had opened a boutique deli and greengrocer: 'Warwick is a much easier place to establish a business than Brisbane. Here you get to know your customers very quickly and you can build your business relationships personally. There's also a lot of interesting fresh foods and produce in the region so I can readily establish a point of difference for my shop over the larger supermarkets' (*Warwick Daily News*, 19 September 2008). A builder in Glen Innes found 'there seems to be less red tape and more efficiency in dealing with the council' (*Glen Innes Celtic Country* 2009). For a motel and restaurant, 'It is great to do business with people who are friendly, caring and go out of their way to assist you ... The University of New England also brings such diversity into the town's culture and has provided us with a great source of good employees' (*Armidale Express*, 16 August 2006). Cafés, motels and new electrical and motor maintenance businesses all figured prominently, as the service sector rather than production benefited. Some sought to develop businesses in new guises. A plumber who had moved to Armidale from Sydney 'found it particularly rewarding working with clients who are committed to making their mark to preserve the planet, by implementing sustainable living solutions in their home. It's what it's all about' (*New England Focus*, August 2009). Small businesses wished to use local produce where feasible. Many of those who had started small businesses (on a much smaller scale than the Burkes) had made the transition to being self-employed, a much heralded transition, usually by developing cafes or small stores, that also implicitly catered to a growing, affluent and discerning population.

Employment

Migrants to the country secured employment relatively quickly, whether they were tradesmen or white-collar workers (the categories most featured). Even if, like the Burkes, wages and salaries might be lower than in the city that was offset by reduced housing and transport costs. Many new residents recognised that wages would be lower in the country but deliberately took a step down in favour of lifestyle. Paul Bennett, a Manager with Blue Mountains Council, found the demands of managing a $80 million budget 'was challenging and often rewarding, but was not conducive to participating in the raising of a young family ... lifestyle and family values were more important than pursuing the "big dollars" that were on offer in the Sydney metropolitan area'. With his wife's family not far away and a brother who had moved to Narromine, he too moved there (*Narromine Shire Council* 2007). For a couple making a similar transition from Melbourne to Rockhampton, to reduce commuting and have more family time, the decision was depicted unusually alliteratively:

> Melbourne is packed with peak-performing power couples pursuing prosperity, like Peta-Anne and Geoff Higgins. When parenthood came along however these professionals paused, prioritised and pulled-out (Rockhampton RDL 2007: 21).

Research had often been done. In an article entitled 'They Came, They Looked, They Stayed' a couple noted 'We thought Oberon had better job prospects when we saw Oberon's low unemployment rates on their web pages' (*Oberon Review*, 4 August 2005).

Professionals claimed to find country employment more rewarding, usually because of greater diversity and more personal involvement: a Glen Innes dentist found 'dentistry in a country town was a lot more rewarding as she performed a broad range of treatments and it was not about mega-profits' while her husband, a doctor, was 'still able to find time for the kids' (*Glen Innes Examiner*, 6 August 2009). In Tamworth, a doctor who enjoyed the 'obvious' benefits of living in the country, more free time and more participation in sport, argued that at a professional level 'the quality of medicine I can practice is superior to what I could do if I was in the city'. Without specialists around opportunities were greater (*Make the Move*, Country Week 2008). Employment had satisfactions beyond salaries.

Affordable Housing

Only exceptional accounts ignored cheap housing, and the consequent savings that enabled various activities to be undertaken, even, as for the O'Connors, having another child. 'It's nice and cheap here which is good – we didn't expect to find a place so quickly' (*Oberon Review*, 3 September 2006). 'You can rent a reasonable newish house in Armidale for the same price range that you could rent a very small older flat in Sydney and the same goes for purchasing a new house' (*Armidale Independent*, 9 August 2006). Units could become houses and small houses become larger. 'They purchased a split level country style residence on 20 acres in town and had some left over after selling their previous home' (*Grenfell Record*, 1 August 2008). For one retired couple, 'Selling up in Sydney gave us the opportunity to buy a home in Gunnedah with enough cash left over for a car and caravan and travel' (*Namoi Valley Independent*, 10 August 2006).

Similar housing would never have been attainable in the city. 'You don't have to be on millions to afford a nice house with a decent garden in a safe environment where you would want to raise a family' (*Narromine Tree Changers*, ms 2007). 'The combination of a mild climate, good work opportunities and much more affordable real estate made the decision much easier ... my friends in Sydney are really struggling, in terms of paying off or buying a home' (*Armidale Independent*, 6 August 2008). 'City living is not much of a lifestyle when you are paying off a million dollar mortgage' (*Armidale Express*, 16 August 2006).

Tranquil Scenery

Peace, beauty and cleanliness abounded, as in Cootamundra where the Jacksons found 'the whole atmosphere in the town is wonderful'. In Armidale 'being able to see the stars at night is magical' (*Armidale Independent*, 9 August 2006). For another couple moving to Armidale was 'a big decision' but

> We got what we'd hoped for, big skies; a safe friendly environment – peace. We only saw tiny slices of the sky in Sydney, and they were more beige than blue. There was constant chaotic noise, with sirens all night, endless traffic and trains rattling by our unit. Now we've found peace ... The lifestyle we found when we came to live in Armidale ... amply makes up for anything we left behind in

Sydney by 100-fold. There's the fresh air, flowers, kangaroos in the back, koalas
... country characters who've never lived anywhere but somewhere like Guyra
all their lives (*Armidale Independent*, 9 August 2006)

'I love the four distinct seasons, I love the heritage and story in each and every
building' (*Yass Tribune*, 6 August 2008). The distinct seasons of the dividing range
attracted others too, including migrants from the semi-tropical northern coast,
where heat and humidity could be excessive. Oberon too was somewhat different:
'We're getting used to the cold still, but we love the countryside, the fresh air and
the forests. It's good having no pollution, no sirens and no traffic as well' (*Oberon
Review*, 10 August 2006). Otherwise climate was rarely mentioned (other than in
terms of air quality) except in the contrasting case of tropical Rockhampton and
coastal north Queensland: 'the lifestyle is laid back and relaxed, which we are
loving after years of fast paced Melbourne, and the climate has been everything
we dreamed of and more' (*Whitsunday News*, 19 September 2008).

For those who chose to live on or beyond the fringes of country towns, peace
was greatly appreciated. In Warwick a couple moved to a 140-acre farm:

He can get to work in 15 stress-free minutes yet his nearest neighbour is out of
sight just over the hill. 'We really appreciate the privacy in Warwick, which just
isn't available in the city. We lived on a standard block in Brisbane. It was far
too noisy for us, so we went looking for "serenity". We found it here. Now the
kids have pet chooks, a pony and room to run about and be kids ... it's all about
lifestyle and enjoying seeing firsthand the kids growing up' (*Warwick Daily
News*, 19 September 2008).

Traffic chaos was universally a thing of the past. 'Sunday driving every day'
(*Oberon Review*, 10 August 2006) and, with 'less traffic, you couldn't get run over
if you tried most of the time' (*Yass Tribune*, 9 August 2006). Peace took various
forms.

The scenery could be appreciated in other ways. 'Warwick enjoys motorbike
riding and therefore found the Oberon scenery beautiful for his outings' (*Oberon
Review*, 7 August 2008). 'Michael Godford's hobby is photography and he says
with Glen Innes district's topography, wildlife and climate "there's no place on
earth like it"' (*Glen Innes Celtic Country*, ms 2009). Beyond regular references to
clean air and rural views and landscapes, few references were otherwise made to
particular pursuits that might be more accessible in the country. Though at Oberon
'there are many activities that both adults and children can enjoy that cannot be
accessed in the city. They regularly go mushrooming in season, take picnics and
drives in the national parks and generally explore the district' (*Oberon Review*,
6 August 2009). Moreover previously urban residents took up broadly similar
occupations in the country. Only rare mentions were made of even horse-riding,
gardening or fishing, and 'growing vegies on the half acre'. Good residents were
relocated urban populations rather than new agriculturalists.

The City in the Country

Not only was it explicit that migration to the country meant no loss of obvious amenities and services such as shopping, health and education, but that the quality of services was at least comparable with that in the city. In Glen Innes 'we have been lucky medically with great doctors and dentists available. We're pleased to have a great working hospital without the long waiting list that goes with living in the city' (*Glen Innes Celtic Country*, ms, 2009). Primary education was similarly lauded.'

Nothing was made of the presence of amenities that might not have been anticipated in smaller towns, although that was sometimes implicit in accounts of delicatessens and cafés. More evidently, as in Cootamundra, 'it has everything we need without the things we don't' (*Cootamundra Herald*, 1 September 2006). Unsurprisingly more sophisticated needs were more likely to be found in larger towns such as Armidale:

> You've got everything you need here. You've got abundant retail and coffee shops. Armidale might have more coffee shops than anywhere else in the world. You can go shopping and find anything. My wife is Japanese and she can find all the ingredients and Japanese foodstuffs she wants. The book stores, art galleries and even the library are fantastic … There's a thriving arts community. I've found more culture in Armidale than Newcastle. The perception in the major cities is that rural and regional centres somehow lack culture, so there's a rural 'cringe' that exists (*Armidale Independent*, 6 August 2008)

In Tamworth it was suggested that towns between 10,000 and 50,000 residents might well be described as 'micropolitan' to recognise their urban qualities. More frequently it was regularly stated that migration to the country meant no loss of access to the city, since it was not far away – even if time was translated into flying time – hence, as for the Burkes, a regular 'city fix' was possible.

The Wards found Narromine 'one of the most central regional locations around. Within 4 or 5 hours by car you can be in Canberra, Sydney, Bourke or on the coast. Or if you choose to fly it's only half an hour to Dubbo and less than an hour to Sydney, all for a very reasonable price' (*Narromine Shire Council*, 2007). 'We're nice and close to Canberra but far enough away to have our own identity' (*Yass Tribune*, 6 August 2009). 'Real estate affordability was one of the reasons to come to Oberon, but also its closeness to Sydney means you can have the best of both worlds … You can walk anywhere here, and there are plenty of shops. And it's only a quick trip to Bathurst" (*Oberon Review*, 7 August 2008). In Stanthorpe, 'We love our quiet tranquil town, yet we are within a few hours drive of Brisbane and the Gold Coast' (*Warwick Daily News*, 19 September 2009). Without exception distances were measured in driving, or in Queensland flying, time and public transport was never referred to.

Occasionally connectivity appeared as a means of 'escape': 'Yass is a good little country town. It's very convenient to all the places you want to go to, such as Sydney and the coast' (*Yass Tribune*, 9 August 2006). Oberon was frequently singled out as a town with good access to other places: 'Oberon's central position to other centres has also been an advantage ... with Goulburn, Lithgow and Bathurst close by. "If you can't find what you need in Oberon, which has not happened to us, but if it does bigger towns are close by" [and] both Canberra and Sydney are easy drives away' (*Oberon Review*, 6 August 2009). In Glen Innes 'Michael and Barbara like to drive to neighbouring towns for dinner, such as Tamworth, a little more than two hours away, or Armidale, an hour' (*Glen Innes Celtic Country* 2009), although that seemed to challenge both service availability and reduced driving times.

Conversely connections meant that relatives and friends could visit the country with ease. 'As the Edwards family discovered it is only a four-hour drive from Sydney and therefore you can get to see you city-bound rellies [relatives] as often as you like, or they can get here without effort' (*Grenfell Record*, 1 August 2008). 'They built a great house on the hill and their families enjoy visiting and experiencing the change in seasons, the peace and quiet and the growing garden' (*Crookwell Gazette*, 7 August 2008). Children of older couples remaining in Sydney found Oberon 'their little retreat' (*Oberon Review*, 7 August 2008).

Distance could be transcended in other ways. Thus, 'though all our children and grandchildren are in Queensland, since Glen Innes has Broadband/Internet we are able to keep in touch with them through modern technology and webcamming'(*Glen Innes Celtic Country*, ms 2009). Shops were similarly accessible. 'We have the internet, so some of the more unusual services that we had access to in Sydney, we can contact them via the net anyway' (*Armidale Independent*, 9 August 2008). Modern technology was otherwise never referred to in a context where the effective extension of broadband to regional Australia has been a contentious issue.

The Bad City ... Babylon

That regional life was a great relief from the city was a constant pervasive theme. Indeed it was a regular theme of Peter Bailey who claimed in an interview in the *New England Focus*: 'I always thought the best view of Sydney was when coming out, through the rear vision mirror as you left the place and came up here' (*New England Focus* 2008). Almost all of those portrayed, like those on the CW website, had moved from larger cities, invariably the state capitals (and Canberra), but usually Sydney, rather than from elsewhere in regional Australia. Just two families had moved from other cities (Newcastle and Townsville), two from overseas and only two from similar sized regional centres. One moved from Kangaroo Valley to Crookwell; the other was a couple who had moved just 40 kilometres from Dubbo to Narromine to purchase a hundred acre block and set up their own motor mechanic business (*Narromine Shire Council* 2007).

Without exception, the majority of those who were moving from Sydney came from western and south-western Sydney, some fearful of the city expanding and engulfing them. Even the Blue Mountains 'are no longer the country destination they were for the city dwellers of yesteryear. They are now the outer suburbs of Sydney's ever-increasing urban sprawl' (*Armidale Independent*, 6 August 2008). At least implicitly they were moving away from something as much as moving to new beginnings.

A couple growing up in south-west Sydney witnessed the 'growth in crime, traffic and pollution ... until one day they looked at each other and said "enough is enough"... the constant pressure of city living, travel and sterility was not for them' (*Cootamundra Herald*, 1 September 2006). 'We had always dreamed of getting out of the city, out of the pollution and overpopulated suburbs of Western Sydney ... it takes time to realise that you can feel happy and safe somewhere' (*Grenfell Record*, 9 August 2009). Similarly new Oberon residents 'have been impressed with the low level of crime in and around Oberon compared to the city' (*Oberon Review*, 6 August 2009). Moreover

> You get numb in the city. You know that you are going to spend a lot of time in traffic to get from A to B, or you are going to see a drug bust or somebody being attacked, you get disturbingly used to all that (*Armidale Independent*, 9 August 2006).

'We did not want to raise our children in the fetid air and grit of the dusty dirty city' (*Cowra Guardian*, 5 August 2009). Most households had similar perspectives.

Some were returning to places they had once been familiar with or seemed redolent of peaceful times in the past. 'Glen Innes offers a quieter lifestyle. It's casual; things just don't happen in a hurry. It reminds me of Grenfell, a small country town near Orange, where I was born and spent my very early years' (*Glen Innes Celtic Country*, ms 2009). The contrasts were evident.

Negativity about the country was conspicuous by its absence, or more than counteracted by the attractions, except in the first special CW edition from Yass where, after complaints about the weather, it did have 'friendly people but bad tasting water' and 'a low crime rate except for last night's two break-ins' (*Yass Tribune*, 9 August 2006). Such aberrant perceptions never reappeared, and every single story focused entirely on the benefits of tree change. Even characteristics that might seem negative could take on new lives. 'The Sheltons have found the climate of Oberon very pleasing, especially the snow. "The cold here is a different cold to Sydney, it's fresher" Robyn said' (*Oberon Review*, 7 August 2009). 'people say Glen Innes is cold in winter but you rug up, light the fire and by 8am it's beautiful' (*Glen Innes Celtic Country* 2009).

Community overwhelmed any initial misgivings. 'We were concerned about leaving behind friends and schools. We needn't have worried as the people of Cowra welcomed all of us. The schools here are excellent and the kids have plenty of activities to fill out the weekends ... As for missing Sydney – we go back for the occasional weekend and see more of it than when we lived there' (*Cowra Guardian*, 5 August 2009). In South Burnett there were hints of challenges, but also opportunities. For a couple returning after some years away:

> It has a wonderful library, there's always something going on here and people are mostly friendly … the locals are open to having a go at something new, new ideas. There's a woman in town now making and selling sushi and people are buying it. This is an area that is growing within itself and becoming more open to different things. We're enjoying being a part of that (South Burnett Regional Council 2008).

In Glen Innes one professional migrant observed that 'you do give up your privacy to an extent, but the big plus was that you become part of the community [especially compared with her parents who had lived in the same house in Sydney for 40 years where] … there was no sense of community' (*Glen Innes Examiner*, 6 August 2009). In the rather exceptional case of Yass, 40 kilometres from Canberra, where significant numbers of residents commuted for work, this meant that 'our area has "outside money" coming in, and there is less of the insular feeling that affects some country towns' (*Yass Tribune*, 6 August 2008). But insularity also had advantages. Even in the tiny remote town of Hughenden, what might seem to be limited retail services meant 'less temptation to spend unnecessarily' (*Hughenden News*, 19 September 2008). Even a hint of negativity was more than cancelled out by other benefits of country life.

Live the Dream?

In 2008 the first of two issues of the magazine *Live the Dream* appeared, a 170-page glossy magazine sponsored by CW, subtitled 'Sea and Tree Change Australia', strongly adhering to and advocating CW's themes (and indeed in both issues it featured CW as 'decentralisation in action'). It contained many stories devoted to the advantages of relocation, with a series of case studies of migrants, quotations from whom follow below. None of the many articles, or the promotional pieces for various regions, were authored. Whether it reached many newsstands is debatable, but it was available at Expos to support more specific documentation.

What gave it distinctiveness was that it covered every state and territory, although by the second issue – a year later – coverage had been reduced to the four eastern states of Tasmania, Victoria, NSW and Queensland, and its size had shrunk to 126 pages. Its philosophy was evident in the first editorial:

> Traditionally, and in the works of some of our greatest poets and artists, the bush has been celebrated as the 'real Australia'. But as our cities have grown into sophisticated global metropolises, regional areas have been marginalised and frequently dismissed as cultural and economic backwaters. In recent years

however the tide has started to turn. Regional Australians still make the journey to the big smoke in search of experience and opportunity, but as they do so they are increasingly passing some heavy traffic moving in the opposite direction. There can be no denying that Australians are being squeezed out of the cities in ever-increasing numbers. Hours that should be spent at home with the family are increasingly spent stuck in traffic. And if the great Australian dream is to own your own home, most young Australians in our capital cities have stopped dreaming long ago. But there is an alternative and many of us are starting to catch on. Modern Australians are starting to return their gaze to the nation's spiritual home, and regional Australia is returning the favour by offering great opportunities in the form of affordable housing, the chance for fast-tracked career progression, and a lifestyle free of the city's frenetic pace. It also offers community spirit in exchange for the city's anonymity (LTD 2008: 4).

These themes were developed further in the pages that followed, which began with invocations to consider moving, emphasised its benefits, partly through the accounts of those who had made the change, and featured a series of 'hotspots' throughout Australia (most of which were large coastal towns and cities, such as Coffs Harbour, Port Stephens and Cairns) that were benefiting from change and which offered particular attractions. (Relocation to such larger coastal centres, with a greater focus on small business development, is not discussed further below). Hotspots were said have high levels of visual amenity, character and heritage, be accessible from a regional centre, have a range of housing stock (including 'rural lifestyle properties') with appreciating values, and 'local products such as wine and gourmet food'. LTD was not averse to offering suggestions for towns that sought to encourage migration, for example through developing festivals: 'If you want to get your town noticed a big annual event can really attract some attention' and the most unlikely festivals, such as the Talbot Yabbies Festival in country Victoria and the Boorowa Irish Woolfest, known for its 'running of the sheep', could attract particular interest (LTD 2008: 22-3). Otherwise parallels with the characteristics suggested by Salt (Chapter 2) were not coincidental. By the time of the second issue LTD had an even greater focus on business and lifestyle, especially movement to farms, and had become rather more like *Country Style*.

A degree of caution was suggested when considering moving to the country. Bernard Salt, described as the 'guru of sea change', advised careful thought and study before moving: 'Don't convert your city property into an upgraded lifestyle property without leaving yourself a financial buffer' and 'rent a house there for a few months' since people 'need to adjust to the change of not having their favourite deli, café or restaurant nearby' (LTD 2008: 11, 12). LTD recommended the necessity of doing the appropriate homework, through developing a checklist of requirements and making a careful study of financial issues and, in its second issue, recommended taking professional advice: 'If fear and uncertainty have left you paralyzed in pursuit of your life changing move, maybe it's time you called in a sea and tree change coach' (LTD 2009: 10). Such 'coaches' provided advice on strategies, finance and destinations, for those who could afford such services, but again emphasised the need for initial exploratory temporary stays and detailed homework. Somewhat surprisingly, LTD also advised on how to deal with snakes.

Getting such basic issues right would engender the real benefits that change offered. One such benefit was a sense of community, and others were remarkably similar to those expressed by good residents elsewhere:

> One of the main advantages of living in the countryside is the sense of community and responsibility among the residents. Rather than feeling anonymous, which can be common in big cities, people actually say hello to you, and they know who you are. They will look up and smile if they are tending their garden (LTD 2008: 16).

Community was readily accessible to everyone. In the words of one who had moved: 'By participating in the community you build really strong social networks and you end up with heaps of friends' (LTD 2008: 33). In regional Victoria 'It feels like all the locals know each other somehow – that they went to school together, or have some kind of relationship' (LTD 2008: 61), a situation that might also be challenging. For many who were returning to regional roots, that was made easier by having relatives there and knowing what to expect.

Time for family was also possible: 'there is more of a chance for the whole family to spend time together and do things as a family, like going to watch the football' (LTD 2008: 17) which also 'provides a crucial social network for the people of [small towns] but also brings much-needed custom to local businesses' (LTD 2009: 119). Children were most likely to be beneficiaries of extra time: 'There is more of a sense of freedom, which means children can grow up in a generally safer environment' (LTD 2008: 17). In the words of one parent, regional Victoria is a

> better place to bring up children because it's community-based, family orientated and the people have good standards... Children have the freedom to safely run, ride and interact, and there is a smaller ratio of teacher-to-child in the school (LTD 2008 : 54).

In Condobolin 'my kids are exposed to experiences – farm life, picnics and fishing along the riverbank – that they would never have had in the city' (LTD 2008: 66). So 'we will stay here and have more kids' (LTD 2008: 33).

Single people could also make the move, though it was not necessarily easy. A 33-year old journalist observed: 'Sure it's harder to meet young people and partner up in your thirties in a smaller place, particularly where everyone knows everyone, but I think there is a growing trend to this sort of lifestyle' (LTD 2008: 95). Another 25-year old physiotherapist in Shepparton, Victoria, was uncertain whether he would stay: 'It all really depends on whether I meet someone ... I definitely think it's a little harder to meet single people in the country' (LTD 2009: 15). Indeed 'expecting to meet a Diver Dan [the sophisticated heart throb of the television series *Sea Change*] would mean that you'll probably be disappointed' (ibid).

Couples who had made the move extolled the benefits, ranging from self-employment to superior health. Sam and Donna Young had left 'Melbourne's peak hour gridlock for the community's country spirit' and established a graphic design business in regional Stawell, that grew into its own office employing three people (LTD 2008: 33). Others had established guest-houses, shops and galleries. A chef who had hesitated to move, not wanting 'to leave the city and then stagnate' found that he could grow 'personally and professionally' even in a town of about 3,000 people (LTD 2008: 61). Even better, there was a positive work ethic in the country: 'In the country there are generally less people and less money, which means people are not afraid to roll up their sleeves to make it happen' (LTD 2008: 62). Technology enabled home business development and online shopping and 'allows people to keep in touch with their family and friends even if they are living far away' (LTD 2008: 17). No disadvantages to business development were mentioned.

Professionals such as doctors were well represented in the stories, moved for the same reasons as others and also found that 'unlike big city hospitals you get to know all the staff ...[and] I've got great flexibility' (LTD 2008: 45). Moreover 'you usually work in multi-purpose facilities with a smaller team so by nature your work is more varied with more autonomy' (LTD 2009: 79). Teachers similarly found that 'because it's a small school there are lots of professional development opportunities for teachers such as training courses' (LTD 2008: 66), while 'It's so easy to make friends with people and you really get to know the parents of your students very quickly. It's an extremely friendly atmosphere' (LTD 2009: 86). And for professionals too there was more family time: 'It's the nicest feeling to get up and walk 500 metres to work or to the kids' school. I've now reclaimed a few hours in my day' (LTD 2009: 78).

Recreational opportunities abounded: 'I play football and golf and ride motorbikes and Donna rides horses' (LTD 2008: 33). Hiking was possible and one couple were now training for a triathlon. The environment too was much closer: 'you become more aware of elements like the wind blowing, the changing colour of the sky, and how the clouds are moving' (LTD 2008: 161). At a more gentle pace: 'I do lots of walking with our three year old son ... Walking is one of the absolute pluses of living in a small town' (LTD 2009: 93). A more outdoor life might benefit health while nutrition too could be transformed; in the smallest places 'there are no takeaway food possibilities, therefore you're almost forced to eat healthily... you don't have a choice but to eat good food because that's all you've got' (LTD 2008: 55) and 'I feel blessed to be ... growing our own vegetables, cooking food from the garden' (LTD 2009: 95).

By contrast the city had inherent disadvantages, mainly traffic, pollution, congestion, crime, anonymity and cost. Houses were relatively cheap in the country.

'We bought our own home within a year – for a third of what an equivalent house would have cost in Sydney. Our mortgage repayments are less than our Sydney rent' (LTD 2008: 66). 'As a 25-year old it would be impossible to even consider buying a house in Sydney, yet last year I was able to buy a brand new duplex here in Narromine' (LTD 2009: 86). 'We have been able to purchase our first home' (LTD 2009: 90).

The country environment contrasted starkly with city life: 'I love being able to go outside at night, look up at the sky and see the stars going on forever' (LTD 2009: 120). 'There are no traffic jams and less pollution means you can breathe clean air. In the city it can feel like you are constantly being bombarded by noise and traffic' (LTD 2008: 17). Here 'there is much less traffic and a welcome absence of road rage' (LTD 2008: 133). 'Even though you don't know it, there's a base hum in the city, right underneath everything. It's like being on stage with a heavy metal band sometimes, because you've got this thudding hum underneath you' (LTD 2009: 52). However some long-distance commuters, mostly recent tree-changers, travelling 90 minutes into Melbourne by train, enjoyed combining urban employment with rural life and found that they made friends and established new social networks with fellow commuters: 'There's about 12 in our little group; we throw ideas at each other for input around our work, celebrate birthdays and have been known to spill out of the train in Seymour [country Victoria] into a restaurant'. They had managed to obtain the 'best of both worlds' without severing ties with either place (LTD 2008, pp. 52-3). Commuting could sometimes be pleasurable.

In further contrast the country provided 'a clean and relaxed environment, which is especially important for our daughter as she grows up' (LTD 2008: 33). 'There's less people, less stress and more time to enjoy what you are doing' (LTD 2008: 55). It was possible, and necessary, to slow down:

> When I lived in the city I always felt like a country girl, but when we moved to the country I realised how much of a city chick I had become. I was used to walking fast and with purpose ... but I hope I have learnt to slow down, smell the roses and take myself less seriously... When you are a blow-in it can be easy to arrive with a mind brimming with new ideas, but even though you may be excited you have to take it slowly. Change is not attractive to everyone (LTD 2009: 92).

Adjustment to country life took time; 'it took a good year to calm down [but] it's because we're artists that we're here. That's why we made the choice to leave the city ... because I don't believe that Sydney is a place for artists anymore. I was always a bit of a hippy. I listened to Joni Mitchell when I was a girl' (LTD 2009: 53). Being

> In a small community we've also had the opportunity to be involved in many worthwhile events that a big city could never have offered. These included judging the 'Lions Youth of the Year' quest, adjudicating school debates, being

a steward at the Country Women's Association cooking competition, being actively involved in the Playgroup Movement, being an active member of the Hay Children's Mobile Service (LTD 2009: 90).

Barely a handful had embraced substantial difference:

Huge fruit trees on my block mean a seasonal bounty of figs and mulberries to harvest, eat and give away. I played two-up with my neighbours on ANZAC Day and I have a real shot at winning 'Best Chick's Ute' at the next Australia Day parade. I experienced the glory and gory of birth when I was midwife for my twin sheep. The locals are poets with profanities. If I wanted to I could go pig shooting. None of this seems very likely in the city (LTD 2008: 71).

The pleasures of pig hunting were utterly ignored elsewhere, and little was made of such activities as horse riding, golf, hiking and fishing.

Changes were not without some disadvantages and many migrants missed something, often friends and family. Some found they needed assistance in easing into a new location. One couple who 'had to settle in to a more conservative lifestyle and a slower pace of life' found that initially 'it was quite difficult to meet people [since] though the local people were quite friendly they already had established friendships and their own lives' hence they formed a Seachange Group that brought in many members and 'served as a support network for newcomers that may feel isolated or lonely in a new community' (LTD 2008: 40). A single woman in outback Queensland had similarly set up a social committee for newcomers (LTD 2008: 113). Another noted 'Sometimes when you move to a small town you can still feel like an outsider even after ten years. But we found that joining clubs was a great way of meeting people. We created a social network by meeting people through a local netball club, through the local high school and also through work contacts' (LTD 2008: 123). Implicitly retirees would thus have had greater difficulty in participating. By contrast 'I initially found the lack of privacy living in a small town confronting. But two years ago when my son was hospitalised with asthma I was touched by the concern and generosity that can only come from a small caring community' (LTD 2009: 90).

Some missed 'being able to grab a great coffee…[and] the retail therapy' (LTD 2008: 55). Food was often mentioned: 'the only thing I miss is the restaurants' (LTD 2008: 134), 'I sometimes miss my gourmet foods [and] the variety of Brisbane delis and markets, but you don't need 22 different types of feta cheese to have a good lifestyle' (LTD 2009: 79). Flies were unwelcome: 'swarms of them. Locals casually brush them off, while I wildly windmill my arms and smack myself like a crazy person' (LTD 2008: 71). A journalist, who had moved from Newtown in inner Sydney only as far as the northern beaches of the city of Wollongong, 80 kilometres to the south, still missed 'the night life, the diversity, the buzz, the shopping, readily available great coffee, my city friends and the chance to report on the constant stream of hard news stories' (LTD 2008: 65).

Multiple advertisements for regional nursing and medical professionals, and for country teachers and lawyers, suggested other things that might be missing. Just one individual recognised that small town life might not suit everyone:

> It makes me sad when I see friends leave Nundle ... [since] we have many friends with people of a similar age and lifestyle. But Nundle and country life is not for everyone. Friends have moved to capital cities and coastal towns for better work prospects and improved access to the services that come with higher populations (LTD 2009: 94).

Friends could be missed who were once seen on a weekly basis: 'Now I only get to see them every year or two. But now we're part of a community and with a booming business and the most beautiful bike trails on our doorstep, there could never be any time for regrets' (LTD 2008: 80). Despite such disadvantages the new migrants in LTD had all made successful transitions.

Partly because the stories lacked individual authors, clichés tended to pile up on clichés, alongside a significant degree of generalisation. Repetition built up a consistent picture of places where family and community were at the core, business development was possible, the environment remained clean and open, there was time for recreation and everyone could benefit from sea or tree change. While it was inherently a good move in the transition to living the dream, it still had to be thought through carefully because places varied.

Rather differently from the local newspapers the promotion of particular places and positive themes by LTD was primarily aimed at professionals and promoted the larger towns – the hotspots – rather than the more remote places where skills were most in deficit. In the advertisements however, remote inland places featured more prominently ('Country teachers broaden their horizons', 'Health care careers in rural and remote Queensland' 'CQ Nurse. Remote Possibilities', 'Are you a GP Looking for a Tree Change?') and the lack of professionals was more than implicit. Thus even within LTD it was evident that certain 'hotspots', mainly coastal, were likely to be the principal places of attraction (especially for middle class professionals) and the beneficiaries of sea change, whereas tree change was somewhat less likely and improbable in remote Australia. Despite some recognition of disadvantages, LTD were advocates for a cause rather than dispassionate observers and analysts.

Rural Realities?

For all that *Live the Dream*, CW, the local media and the many councils who attended the Expos sought to paint a positive image of regional Australia, and move away from the bleakness of dust, dead sheep and droughts, many depictions elsewhere are of a regional Australia that experiences problems. A persistent drought has gripped large parts of rural Australia for much of the present century, even if its impacts have been less evident in the larger regional towns, alongside the impacts of recession, the GFC and fluctuating global prices on core rural industries, and intermittent and occasionally devastating bushfires. Some regional towns have social stigma attached to them, at least in the limited knowledge of many urban residents. Some, especially in western NSW, were experiencing another version of 'white flight', as white residents sought to move away from what had sometimes become troubled towns, marked by high levels of Aboriginal social deprivation, unemployment and intermittent crime (and in a region where agriculture was experiencing prolonged crisis).

Such issues have long been recognised. What was true of Moree towards the end of the 1990s has been true of many other regional centres in recent years. As the mayor, Greg Jones, then stated:

> We are facing difficulties with the downturn in the agricultural industry. We got little wheat last year because of the drought and we had no water for irrigation. Over the past two years there has been no agricultural income, so no disposable dollars have been coming into town and that has affected business and local employment. A lot of families have left to look for work elsewhere. When you have downturns like that you have unemployment and that means people being bored, which can lead to problems like vandalism. During the recession a few years ago Moree wasn't affected but the last drought has affected all of the agricultural industries. This may be the third year running that the area will have no water allocation from the river. It's going to be disastrous for a lot of people and for businesses. If you walk down the main street you'll see a number of shops closed down. The smaller family businesses are being taken over by the bigger corporations. This ... means the face of country towns is changing. We are getting shops like Fair Dinkum Bargains and Sheer Madness and Crazy Prices, that level of shopping. That's all that people can afford (1997: 22).

While Moree is more remote than most regional towns in NSW and Queensland, and droughts have tended to be more devastating further west, other towns have experienced crises and nearby Inverell had similar problems:

> Inverell, like many other medium sized regional/sub-regional centres is under significant pressure, despite some bounce back after the recent drought. Long term difficulties with the agricultural sector accompanied by a lessened dependence on on-site labour have weakened the economic base of the region. In addition, the substantial withdrawal of government, social and community services with the attendant loss of government jobs has removed another major economic underpinning of the town. These processes, operating together, are largely responsible for the static and/or declining population of Inverell ... (Witherby 2001: 202).

Similar problems have affected many smaller towns at different times in the past couple of decades with a harmful effect on morale (Smailes 1997). Lack of social services has posed problems in many regional areas hence the frequent advertisements in *Live the Dream* and elsewhere, for teachers, doctors and nurses. In Hay (NSW), for example, the town's hospital was 'grossly understaffed' and there was no maternity ward; the town's only full-time doctor was leaving in 2006 and no replacement was imminent (Grigg 2006). More generally access to health care has worsened. Suicide rates are higher in regional Australia than in large urban areas (Caldwell et al. 2004), physical and mental health are both worse than in urban areas (Strong et al. 1998, Rawsthorne et al. 2009), divorce rates can be very high, domestic violence is not uncommon (Wendt 2009), crime has multiple dimensions (Barclay et al. 2007) and poverty has been increasing in 'declining country towns' (Baum et al. 2005: 1).

Yet at the same time many problems are quite localised and tend to be experienced most dramatically, and most visually, in rural areas where droughts especially are not simply powerful brakes on agricultural development (with impacts on the service sectors of nearby small towns) but have become almost the leitmotif of television coverage of rural Australia. On television at least rural Australia all too often appears to be in crisis. It is widely believed, certainly among CW's hierarchy, that television lingers unduly and visually on depression, hence the 'dead sheep in the dam', rather than on more positive aspects of regional life. At a distance all news is bad news. Since most Australians live in large coastal cities media coverage of rural and regional issues is quite limited (especially in prime time slots) hence experience of regions is largely confined to dramatic events.

Drought is a recurrent problem, and of inevitably uncertain duration and periodicity, while it has increasingly been linked to climate change towards a hotter Australia, and the equally uncertain politics of allocation of licences to river water for irrigation and other uses. Salinity, over-grazing and rapid clearance threaten some areas. After eight years of drought, late in 2009, the Lachlan River, flowing through the towns of Condobolin, Forbes and Cowra, was expected to dry up, so that neither Cowra nor Forbes (both towns of over 6,500 people) were expected to have any water: 'thousands of households will have to truck water in to live in their houses or they will have to walk away' (Wilkinson and Cubby 2009). The much larger town of Dalby (Queensland) was within eight hours of running out of water late in 2009. Under the title 'Endangered Species. Australia's Disappearing Country Towns', the usually austere *Australian Financial Review* noted in 2006 that 'the future of many towns is no longer assured as water shortages finally bite' since paradoxically more water for the rivers, as 'environmental flows', meant less for irrigation. Inland towns such as Bourke and Hay, dependent on agriculture, were threatened since irrigated land is valued at roughly ten times that of non-irrigated land which would become marginal grazing country, generating smaller profits and much less employment (Grigg 2006). Agriculturalists and conservationists have come into conflict at different scales and in many different places, while which towns may disappear and why cannot be foreseen, merely adding to rural uncertainty. Week after week such issues are debated in the national newspaper, *The Land*, providing a very different sense of rural and regional Australia than *Country Style* or *Live the Dream*.

While CW may rail about biased and blinkered perceptions and misconceptions of regional Australia from urban critics who rarely travel west of the ranges, and there is good reason to rail, multiple challenges confront regional Australia which account for the need for and role of CW. The challenges that face agriculture affect all regional services and the vibrancy of small towns, most of which depend in some way on agriculture. Yet agriculture is far from being the only livelihood of regional Australia; mining has experienced intermittent booms, and larger towns especially have a diversified economy that can withstand agricultural stress, and are thus more likely to prosper.

Making the Change

Like LTD, many new residents were not averse to offering homilies, and recommending tree change to others, some of which recognised that planning was necessary and adjustments had to be made. 'A sense of humour is vital, whether you're here or somewhere else. You know that life in a smaller place will be different – if you wanted the same as what you had, why move?' (Rockhampton RDL 2007: 36). Retired people were offered additional advice: 'If you want to become part of a new community you have to make an effort but once you do this you'll find new interests and just as important new friendships' (*Singleton Argus*, 8 August 2006). For potential business developers in Rockhampton: 'Do your research thoroughly. This may take some time. I took three years to come to the decision – but it's worth the effort... make sure you have your spouse and family committed or at least open to giving it a go. It's important to get out and meet people. Try to hook up with relevant groups' (Rockhampton RDL 2007: 12).

In some particularly small and remote centres a greater degree of advice was offered. In Hughenden, a tiny north Queensland town of some 1,150 people, the Chamber of Commerce advised, in what appeared to be a version of the ten commandments (or the desiderata):

> Come and enjoy what is on offer but without preconceived expectations – have an open mind.
> Be prepared to mix with people and share your ideas.
> Leave your city attitudes behind and come and try it.
> Join in with community activities.
> Give yourself time to adjust and be accepted.
> Just be yourself; don't try to keep up with the Joneses.
> Remember one only gets out of something what one puts into it.
> Be prepared to take time to settle in before venturing away for your entertainment.

> Shop locally – believe in local business people.
> Enjoy the personal service – traders will respond to your requests,

which was rounded out with the 'final piece of advice' – 'you'll never never know if you never never go' (*Hughenden News*, 19 September 2008).

However much research or even caution was required, invariably there was really only one conclusion. 'If anyone has any reservations about making the move don't hesitate – the air is clean, the people are friendly, the houses are cheap' (*Cowra Guardian*, 5 August 2009). 'Moving to Parkes has given us the type of family opportunities we were missing out on in Sydney. I would certainly recommend it to anyone as a viable option. The cost of living is cheaper, life is less hectic and you get to say G'day to so many people in any single day who all respond with a smile on their face' (*Parkes Champion-Post*, 7 August 2009). 'People always say "I wish I could do what you have done" but they think it's not an option … Some people think they can't handle not having a Starbucks around the corner' (*Oberon Review*, 7 August 2008). 'The Gibsons predict that as more coastal dwellers get weary of the sea change and accompanying squeeze, they'll look to places like Wondai to live and work' (South Burnett Regional Council, 2008). Most assumed that since they had made the move others would recognise the wisdom of this, and were prevailing on former neighbours and friends to make a similar move. The 'good residents' offered shared success.

Good Citizens and Good Country

The stories, and pictures accompanying them, highlight successful adaptation to small town life. While these appear as stereotypes, or at least carefully chosen households, certain people may be better migrants. Despite the diversity of destinations, what constituted a good resident was remarkably homogeneous. Good residents were nuclear families, where one or more parents were employable or could set up businesses, valued community and were willing to participate in it and had no esoteric demands that regional services could not support. Retired couples were almost as acceptable. Diversity was not totally absent as gay couples and singles could migrate. They had come for a better lifestyle that involved more family time, recreation, clean air and open spaces, distant from increasingly congested and polluted cities. They had moved into a community where they would be welcome.

There was therefore a rather bland uniformity, which might have been what many were seeking, and a retrospective vision of a past time of older values, a more local space and place, where children could be children, crime was absent, abundant greenery existed and walking to work was feasible. Families and community were the leitmotif of lifestyle as in Hughenden:

The feeling of community is especially important. We all like to feel we belong. We all like to feel we have a place where our contribution is valued. We all like to feel useful in some way to our society. It's easier to feel that way in a country town rather than the city ... [in] a place where people matter (*Hughenden News*, 19 September 2008)

That family life might occasionally be stultifying, a quiet life be soporific, 'family time' be synonymous with the lack of nightlife, quick journeys to work coincide with lack of public transport, or a greater focus on cooking imply a lack of restaurants had no place here. In one small Queensland town, the Tourism Manager observed: 'There are a lot of activities for families, young people, married and retired people. We have 11 pubs'. Rose coloured (or perhaps green tinted) glasses abounded. Even the smallest rural towns were viable, vibrant places (not places of crisis, loss and problems, of drought, fire or depressed prices, or boredom).

Previous relationships with the country eased migration, made it more probable and more likely to be successful. As one editorial noted 'It is not like a foreign country, with foreign ways as it is to some city people' (*Yass Tribune*, 6 August 2008), a somewhat divergent theme from the rest of the special issue which encouraged city people to move to the country, and hinted that no great effort was involved. There was an implicit sense that people would want to live in rural areas given knowledge and the chance; as elsewhere 'Living and acting within smaller units is seen as corresponding with human nature' (Cruickshank et al. 2009: 80). The rural thus became 'a less hurried lifestyle where people follow the seasons rather than the stock market, where they have more time for one another and exist in a more organic community where people have a place and an authentic role. The countryside has become the refuge from modernity' (Short 1991: 34). Similar constructions of the countryside as a place of refuge, of old-fashioned values and good people, exist in the white English countryside (Neal 2009) and the migration of British residents to Lot in south-west France, where 'rural France is presented as the rural idyll, characterised as the Britain of 50 years ago, offering a way of living that the migrants believe is no longer available to them back in Britain' (Benson 2009: 123). So too in regional Australia. Beyond that, as in Norway, 'a decentralised settlement pattern is seen as an important part of national tradition and cultural heritage, and is therefore considered crucial for the building and rebuilding of national identity' (Cruickshank et al. 2009: 81). Rurality offered more authentic values: countrymindedness still existed.

More than two decades ago James wrote of rural NSW that 'the meagreness of rural existence' imposed by isolation, inadequate transport facilities and inferior education, 'may be masked or made bearable by the belief that country life is simpler and morally better than city life'. So that women who listed the privations of country life in the early 1980s also commented: 'It's a great life', 'outdoor life is a great way to bring up kids', 'I would never want to live in the city' and so on' (James 1989: 77). Much has changed but a distinct morality, even a moral economy, is at the core of the demand for and depiction of the good migrants and their role in coutrymindedness. That discourse is informed by dualisms. Yet the economic circumstances that brought many newcomers to the regions have been muted in post-migration rationalisations and discourses of a beneficent countryside. Even in the idealised accounts of the successful newcomers there were underlying hints that many had moved for pragmatic reasons where a new lifestyle was a fortunate adjunct.

Chapter 9

Living the Dream? A Retrospective

Many people who make the move from city to country have country in their blood. Perhaps they were born in the country, grew up with wide open spaces, or spent happy holidays visiting uncles and aunties who remained in regional areas while another branch of the family tried their luck in the big smoke. Such people ... have country on their radar.

Yass Tribune, 6 August 2008

'The best of city and country living'

Signboard 25 kilometres outside Warwick, Queensland

In recent decades almost any form of 'rural revival' would have been welcome in the more remote areas of most developed countries, as regional populations have aged and declined and rural economies been hollowed out or, more occasionally, where resource booms required labour. Yet efforts to directly stimulate revival by encouraging urban–rural migration have been virtually non-existent. Only in Australia, and somewhat differently in Sweden, has direct marketing seemingly been attempted. Elsewhere place branding has centred on tourism and business development. In many different social and geographical contexts, the majority of the literature on urban–rural migration, alongside media narratives of mobility and the themes of popular television series, has largely presented and interpreted existing urban–rural migration as a broadly middle class phenomenon. Much counterurbanisation in Australia does fit this pattern, while classic stories of middle-class migration driven by constructions of a rural idyll and lifestyle preferences, have become 'orthodoxies of counterurbanisation' (Halfacree 2001). In inland Australia such orthodoxies are shattered.

As the last two chapters demonstrated there is a disjuncture between the structure and process of migration to two small NSW towns, which is extremely varied and with a range of outcomes, not all of which have been positive (Chapter 7), and the idealised good residents, often middle class, selectively culled from this diversity, where positive voices and experiences dominate and whose transition to small town life has been remarkably harmonious (Chapter 8). Certain groups are not targeted in the literature or represented as good country citizens, and the magazine *Live the Dream* exclusively focuses on middle-class professionals. Image and reality are some distance apart, and that reality in regional Australia is rather prosaic. Even in just Glen Innes and Oberon there is a massive variation in who the newcomers are; migration has come from a range of locations, nearby but even overseas, for multiple reasons and into diverse new lifestyles. Such diversity is quite usual (e.g. Bolton and Chalkley 1990, Curry et al. 2001), and here as elsewhere newcomers cannot easily be categorised (e.g. Escribano 2007, Osti 2010). Despite that diversity the rationale for migration is more straightforward.

While improved lifestyles and better work-life balance embellish the content both of marketing (Chapter 4) and discourses of change (Chapter 8), underlying almost all moves is the availability of housing and employment. Lifestyle cannot exist in isolation. Where lifestyles have changed they largely represent a shift towards family and leisure, rather than an embrace of new cultural values or agricultural lives. Tree change is no radical break from the recent past, and definitely no return to an agrarian society. The new country is a 'hybrid country'.

Country Week

While this book has focused, like others before it, on 'uni-directional flows of people to rural areas' (Milbourne 2007: 384), it has sought to add additional dimensions to counter-urbanisation by moving away from middle-class orthodoxies and focusing on marketing and branding. The concept of an annual city-based Expo, as an institutional strategy, involving centralised and localised place-marketing strategies that are both collaborative and competitive, is unique. Moreover the Expos are not directed exclusively at middle class migrants, usually assumed to be the sea and tree changers, but at workers of all kinds. Various messages are specifically targeted at people who may find living in large cities challenging, particularly in relation to the cost of housing and the strains of commuting, though Country Week (CW) is certainly not aimed at welfare migration.

The Expos are innovative engagements between urban residents and rural communities, that cross many perceptual divides between city and country, and encourage direct contact between people rather than engagement mediated by newspapers, radio and television. The numbers attending the Expos offer clear evidence of the continued interest in moving to rural areas: neither tree change nor sea change are losing momentum, but CW offers a new strand in counterurbanisation. That new strand is the deliberate search for migrants, who are also workers, hopefully with families, who might be lured by lower house prices and employment opportunities rather than the somewhat intangible lifestyle that is held to characterise sea change.

For much of the twenty-first century, CW has existed to encourage multiple forms of migration, not only recognising the diversity of urban–rural migration, but realising the need for it, and especially the need for tradespeople, unlike earlier processes of rural gentrification and counter-urbanisation, subsumed under the notion of 'sea change'. But CW, with its slogan 'Live the Dream', has sought to have its cake and eat it – in favour of young families and trades-people, and therefore counter the 'orthodoxies', but encouraging and advertising a middle class migration driven by lifestyle choices, so evident in the imagery. The job boards at CW, with their focus on trades, and the housing advertisements (that emphasise affordability) are somewhat different from the glamour and obvious

played in determining our prosperity, our values, our 'place' in the landscape and the myths we have woven about our past, present and future' (Waterhouse 2002: 99), but nostalgia and memory play no part in contemporary politics and regional economic development. The distancing of agriculture from the lived experience of urban residents has meant that a new concept of 'country' has had to be created to attract migrants from the city. While CW subscribes to many of the tenets of countrymindedness (restoring the city-rural population balance, encouraging rural life, with a distinct anti-urban bias), it is sufficiently astute to recognise both that urban residents may not share these ideals and that urban–rural migration weakens such lingering notions: a more contemporary country has emerged.

CW consequently blurs any urban–rural divide. The regular attendance of Canberra, Australia's inland capital of 350,000 people, both highlights the simplicity of any geographical construction of an urban coast and rural interior, and indicates the challenges to migration inland. Places that are part of an extended peri-urban area within relatively easy range of cities have crucial advantages, combining accessibility to the city and some form of rural lifestyle. A short distance from Sydney or Brisbane was invaluable. Nearby larger towns, such as Orange and Bathurst, had no need to attend. Bernard Salt summarised the nature of new residents of such places: 'they are urban people downsizing from the city but still with an umbilical cord – so they can go back if it gets too scary' (quoted in Harley and Phillips 2006: 20). But that is only possible for some; for many selling urban property may mean burning bridges with metropolitan life.

Locations matter. Not every country town experiences such diffuse urbanisation. Beyond the umbilical cord, beyond about three hours from the city, another regional Australia exists. As various visitors observed, the oft-repeated three-hour drive of sea change (Gurran 2008) or the 250 kilometres of tree change (Salt 2009b) was a major deterrent: more than a step too far. And that excludes vast areas of inland Australia: an Antipodean version of the 'deep rural' where there are few activities and services that could be constructed as 'urban', and few attractions for city people contemplating moving to the country. This 'deep rural' is the location of a multitude of declining towns like Nimmitabel, Warialda and Ariah Park. Small towns, bordering on ghost towns, have borne the brunt of regional decline, where boarded up farms, shuttered shops and closed services are far from the idyllic countryside of CW and council brochures. This is the part of regional Australia that is 'doing it tough'.

Such small and remote places most need a boost from the Expos. Such a boost may be slight but, in a context of stagnation, the ability to retain a school teacher (and their family), because one or two new families have moved to the area, is a significant achievement. Performing well at the Expo and doing the necessary follow-up cannot overcome the structural issues that have contributed to decline in particular locations, but it may limit the harm. But, not only do these small and declining settlements find it hardest to attend the Expos, due to resource constraints, when present they are the least known: 'spellcheck towns' that have to locate themselves on a map, because they are not on the mental maps of visitors, and crowded out by larger towns with more services and more effective presentations.

With most Expo visitors seeking to move to the coast or, failing that, to medium sized inland towns, thus furthering the growth of the regional cities, the more desperate plight of small remote places from Ariah Park to Thargomindah can be overlooked. Small towns like Walcha and Cumnock in the penumbra of Armidale and Orange respectively simply had no opportunity for separate representation. Rather larger towns, like Warialda or even Moree, or Narrabri and Narromine before them, could not justify continuing to be there without more obvious positive outcomes, and simply dropped out. Just as farmers may be paid to leave unprofitable farms, dwindling towns in their vicinity may gradually be abandoned.

City Style – Country Heart

Most Expo visitors are considering moving away from the city, not necessarily out of any broad dislike for urban areas, other than in rhetorical moments, but because of housing costs, and the underlying promise of a better life that savings may enable. Appropriate employment for most is essential. Countrymindedness may be an important concept for CW, and thus its focus on lifestyle, but there is little 'countrymindedness' for most potential migrants, just a pragmatic desire for adequate and accessible services and amenities – schools, hospitals and shops – to meet straightforward family needs: not dreams but shrewd and realistic economic decisions. Consequently the 'sponge cities', the better known and larger 'hotspots', many on the coast, and the inland cities that have no need of Expos, few far from metropolitan Australia, are much the most attractive destinations. Location, size and services all matter.

The larger towns, the 'top places', are most attractive because of the range of possibilities they offer, and rather than promoting the country and portraying large cities negatively, councils such as Tamworth encouraged migration using slogans like 'city style-country heart'. Cootamundra redefined the rural as 'new country'. CW itself stressed: 'don't make the mistake of thinking of country and regional areas as "the bush"'. Whatever rurality might have been, it acquired urban attributes. And visitors responded: Oberon all too often was 'too small' but 'We like Tamworth because it's like Sydney only smaller'. In the marketing strategies of larger towns, the new tree change was rarely about the country at all, but simply a means of gaining access to adequate facilities in a reasonably pleasant environment. The recent addition of 'Regional Living' to what was formerly Country Week, was no subtle change but recognised pragmatic shifts from the landscape orientation and tourist focus of earlier years. As one migrant 'forced' to move from Sydney to Orange, when his job was decentralised, observed: 'Now that I live here I think it's great … It's like a Sydney suburb stuck in the middle of the country' (quoted in Pryor and Lewis 2004). New country, and hybrid country, are effectively creating 'urbs in rure'.

A repeated slogan was 'move out of out of town but not out of touch'. Selling points were often the distance to Sydney or Brisbane and the ability to return. New residents may have left the city but the city had not left them. Umbilical cords remained attached. As one good resident in the Upper Hunter Valley stated: 'while we have that [country] lifestyle we are still not far from Newcastle or Sydney to visit … You can still work all week here and go to a show or harbour cruise in Sydney on the weekend. It is enjoyable to visit the city but it is always enjoyable to get back home and have that relaxing lifestyle' (*Muswellbrook Chronicle*, 8 August 2009). Even such good new residents as the Burkes needed a 'city fix', others missed 'retail therapy', McDonald's marked progress and Junee offered all manner of urban attractions. Moree advised that flying back to the city was cheap and feasible: the 'best of both worlds' was essential. This new country might even be seen as 'cappuccino country'; CW itself mentioned a 'cappuccino index', while Junee and other towns used coffee as their symbol of urban comparability. Country towns thus experienced 'domestication by the cappuccino' (Zukin 1995), as metaphor and reality, as new consumption and as a shift to services. What Salt has called 'a bit of inner city *uber chic* culture [has been] transported to the countryside' (quoted in The Australian, 19 September 2007). Hay bales no longer decorated the floor at the Expo, alpacas were irrelevant and the Country Women's Association had gone. New migrants were boosting the places that are most like cities, especially those closest to 'home', and turning them increasingly into small cities: taking the city to the country in expanding micropolitan centres.

Basic Needs

Anticipations of an improved lifestyle underpin migration, but improvement comes from having a job and a house, in a place with reasonable facilities and of manageable size. As one council observed: 'We all do the five minutes to work, the kids are safe on the street thing, but definite employment and housing are the winners'. That scenario is widespread. In rural Devon in England, compared with housing and employment, 'only two non-economic factors' were of any significance in moving and the principal one was family connections (Bolton and Chalker 1990). In Sweden only environmental factors rivaled employment and housing (Lundholm et al. 2004, Hjort and Malmberg 2006), and in Scotland employment and housing were pre-eminent (Stockdale 2010). While migrants often preferred to emphasise lifestyle, amenity or personal connections, all factors in decisions to move, lifestyle played a limited role in the actual choice of location. As much as anything else, a better lifestyle came from being close to familiar places and people, and having time to appreciate them.

Generally however new migrants have not moved far, choosing to be reasonably close to where they came from and preferring a larger town. The destinations that most Expo visitors preferred were remarkably similar to the coastal 'population turnaround' regions that Burnley and Murphy (2004) identified as the destinations of (mainly) sea change movers, thereby doing nothing to rebalance country populations. That is unsurprising: 'it is well known that most migrants do not move far and distance is generally the most important factor explaining the number of migrants moving to distant places' (Niedomysl 2007: 702). Few newcomers in Glen Innes or Oberon were dreaming, while some saw themselves as 'economic refugees', not quite where they wanted to be but finance precluded cities or sea change. Despite improved infrastructure and telecommunications providing superior electronic and physical linkages, the longevity of kin and valued friendships deter distant migration to small and unfamiliar places.

Many of those who had made successful transitions to towns such as Oberon and Glen Innes were returning: regional life held fewer uncertainties, kin were there and myths of idyllic places and pastoral visions had no place in their decision-making. Despite the metropolitan target-market of CW, many people moving into small towns were moving within regional Australia rather than experiencing a tree change conversion from the metropolis. Others spent years 'winding down' through investment properties or second homes, until families were again small enough to minimise complex decision-making or another catalyst led to a move. Smaller towns like Oberon best met the needs of people at particular stages of life, childrearing and retirement, but the superior services of larger towns offered more. Few of those who came from Oberon ever anticipated return (Brown 2006) but in the end it met certain needs, for families or older people. This meant both a limited rejuvenation but also something of a 'greying' of the smaller towns, sometimes accentuating the problems of an already aging population. Younger families sought larger towns.

The stories that migrants tell and that are glorified in the press rarely reflected the diversity yet simultaneous simplicity of migration. People moved around within rural and regional Australia often because they had to: public servants were posted to distant places with the prospect of future promotion, ailing family members and partners required propinquity, employment, housing and family relations were central for most, and lifestyle improvements a bonus. A few were effectively forced out by city costs of living but very few were 'welfare migrants' desperately searching for any place cheaper than the city where they might simply get by. While most enjoyed the peacefulness of regional life, new residents, especially those urbanites attending the Expos, were not in search of 'alternative' lifestyles. A couple from an alternative farm who touted their own sustainable self-sufficient way of life at the 2008 Expo had little resonance for CW visitors, and did not return in the following year. 'Rentafarmhouse' had limited appeal, because of rural isolation and the necessity to be a handyman. Few new residents adopted any distinctly 'country' activities, even country walks or creating vegetable gardens, and even fewer took up agricultural employment. There was no 'big shift' but merely many diverse decisions to relocate to meet basic needs and hopefully improve lifestyle, in a place that was not too unfamiliar.

Yet for all that there is a persistent retrospective vision in much of the rhetoric of CW and the councils – from house prices and open spaces and presumed crimelessness and the absent stress of older times, to versions of 'white flight', the slow pace of life and the stability of changelessness. This conjures up discourses of nostalgia and the 'slow life' – a broader variant of slow food, that cherishes both heritage and timelessness. Ironically almost all councils were at CW because they sought to speed up and modernise the process of change, by bringing in new younger residents, expanding businesses, filling employment vacancies, encouraging the growth of schools and getting their regions on the move. Only the pace of this change, with some smaller councils stating that they did not want a deluge of immigration, differentiated places rather than the nature of change itself.

Marketing Rural Revival?

Marketing place is difficult and cannot be achieved instantaneously. The drop-out rate of mainly remote councils suggests that sometimes it cannot be achieved at all. Migration takes time and careful consideration, unless prompted by job transfers or facilitated by family connections. Many Expo visitors became frequent attendees. Even more importantly, migration is often underpinned by structural issues that provide economic security, and CW is not likely to significantly contribute to that given the relative importance of commodity prices, government planning, corporate investment decisions, wage bargaining and so on. Place marketing can be effective but, as in Sweden, usually only if people were already contemplating a move (Niedomysl 2007) and needed a catalyst, a jolt or at least a degree of encouragement. The Expos could be catalysts, but most people migrated to regional Australia without CW. Nonetheless the Expos had at least as important a role in boosting internal images, even in declining areas, and in stimulating tourism.

Marketing increasingly focused on what some councils saw as the 'four points': housing, employment, services and location, but closely related to these were some notions of lifestyle, that combined amenity and community and a degree of anti-urbanism. Within such common themes distinguishing particular places was never easy, so that location became one of the most important means of differentiation, and for nearby places Open Days and Festivals gave added value.

Regional Australia is neither utopia, nor a polar opposite of some dystopian urban life, but holds challenges, ranging from limited access to certain services to establishing new social relations. Established residents of small towns may be conservative, sometimes resentful of changes and those who bring them, inhabiting a regional Australia that may still be perceived as agricultural rather than linked to the 'new' employment that draws in new residents, and which may threaten established ones. Newcomers are not always accepted willingly, prices are higher for some goods, and other goods are scarce. Choice can be limited; as in Narromine there may be 'only one of everything'. Urban residents who have no prior experience in living in rural areas are justifiably cautious about making a move. If kin are distant, and finding a place in a new social and economic context is challenging, daily life is fundamentally different. For both established residents and new arrivals, change and compromise came slowly.

While CW cannot, and could not be expected to, create a rural revival that would redress population imbalances in Australia, it provided information and offered potential migrants a choice about regional living through an instant virtual tour. The Expos enabled country towns to connect directly with urban residents, and encouraged local governments, the media and small businesses to 'work together'. CW could smooth the process of migration for those potential migrants who might otherwise have never got beyond the 'thinking about it' stage.

No single weekend event, however successful, could possibly transform regional Australia, or significantly influence counter-urbanisation, in the face of national and international market forces and the relative ignorance of regional Australia by so many urban residents (epitomised in questions about running water and electricity). Reducing metropolitan congestion is even more improbable. Counteracting lack of knowledge is a small but vital role, merely a first step towards greater familiarity with 'the bush', and putting places on the map. But seeking rural revival is a worthwhile goal since it has the potential to improve the lives of many people currently living in regional areas, and the lives of people yet to migrate there. But, as Costello (2009: 219) concluded of analogous 'tree change' in Victoria, 'this relatively small influx of people to rural locations for lifestyle gains is not likely to result in widespread rural growth across Australia'.

Although CW plays a small, sometimes lonely, part in contributing to knowledge and interest in regional Australia, the forces of nature that brought the devastating bush fires to Victoria early in 2009 (when more than a hundred people died), and intermittent droughts, all featured prominently in the national press, are constant reminders of challenges that are simply not present in metropolitan Australia. (Ironically the second issue of *Live the Dream* went to press a short time after the Victorian fires with a cover photo and story of a newly constructed, sustainable 'eco treehouse' destroyed in the fire. Its editorial was devoted to the need to ensure that the fire would contribute to removing all divisions between the city and the bush in order not 'to dampen the dreams'). Somewhere between the worst case scenarios of droughts and bushfires and the success stories of the newcomers (Chapter 8) there are distinct and attractive regional opportunities.

And, as even the two small towns of Glen Innes and Oberon well demonstrate, if tree change is the movement from large coastal capitals to small country towns it is still diverse in people, influences and outcomes, all of which preclude easy characterisation. There was certainly no 'collective identity' of 'movers' (Halfacree 1997) and no means of distinguishing them from established residents in a simple dichotomy. In some respects it is easier to ask who were not migrants? In these towns at least there was merely a relative absence of very young families, white-collar workers and 'high-end' professionals (other than the few who had retired). The more educated rarely stray far from the peri-urban zone (cf. Hjort and

Malmberg 2006). More generally country towns may not be for everyone; youth, the highly skilled, singles, gays and 'ethnic' Australians may appear to have a limited place. Aborigines are there but often ignored. Regional Australia is quite conservative.

Despite the success stories of urban–rural migration trumpeted each year at the Expos, the population of inland Australia continues to decline relative to capital cities and coastal locations, and in several instances it is declining in absolute terms in the towns and regions of inland Australia. The growth of large cities is not accidental – there are many attractions for people in cities. Ironically, such attractions are now contributing to micropolitan growth, and to new structures of uneven development in regional Australia. Sponges thrive; ghost towns emerge. Generations of Australians have never lived in rural areas, and, despite *Live The Dream*, rural Australia is not their 'spiritual home' (p. 157) and nor do they have 'country in their blood' or 'on their radar'. Yet, pragmatically, it can become the home of more urban Australians who find aspects of big city life to be challenging, uncomfortable and costly. But economic growth is crucial and it will boost the more desirable regional towns. Not all of regional Australia can possibly grow. Remote areas and the smallest towns will continue to struggle within a market economy. Instead of perpetuating a lifestyle-oriented myth, the practice of urban–rural migration highlights the importance for many potential migrants of mundane but basic considerations such as housing, employment, education, medical facilities and family connections. In a country of dreamtimes, hard realities and practical dreams of work, of home and of family are shaping a new, uneven and hybrid country.

Bibliography

Aitkin, D. 1972. *The Country Party in New South Wales*. ANU Press, Canberra.

Aitkin, D. 1985 'Country-mindedness': The spread of an idea. *Australian Cultural History*, 4, 34-41.

Alston, M. 2004. 'You don't want to be a check-out chick all your life': The outmigration of young people from Australia's small rural towns. *Australian Journal of Social Issues*, 39, 299-313.

Argent, N. 2008. Perceived density, social interaction and morale in New South Wales rural communities. *Journal of Rural Studies*, 24, 245-61.

Argent, N., Rolley, F. and Walmsley, D.J. 2008. The Sponge City Hypothesis: does it hold water? *Australian Geographer*, 39, 109-30.

Argent, N. and Walmsley, J. 2008. Rural youth migration trends in Australia: An overview of recent trends and two inland case studies. *Geographical Research*, 46, 139-52.

Australian Bureau of Statistics 2007. *Regional Population Growth, Australia, 2006-07*. ABS, Canberra.

Australian Bureau of Statistics 2008. *Regional Population Growth, Australia 2007-08*. ABS, Canberra.

Baldacchino, G. 2008. Population dynamics from peripheral regions: A North Atlantic perspective, *European Journal of Spatial Development*, 27.

Barclay, E., Donnermeyer, J., Scott, J. and Hogg, R. (eds) 2007. *Crime in Rural Australia*. Federation Press, Sydney.

Baum, S., O'Connor, K. and Stimson, R. 2005. *Fault Lines Exposed. Advantage and Disadvantage across Australia's Settlement System*. Monash University ePress, Melbourne.

Bauman, Z. 1995. Searching for a centre that holds, in *Global Modernities*. Edited by Featherstone, M., Lash, S. and Robertson, R., Sage, London, 140-54.

Beer, A. 2006. Regionalism and economic development: Achieving an efficient framework, in *Federalism and Regionalism in Australia: New Approaches, New Institutions?* Edited by Brown, A. and Bellamy, J., ANU E. Press, Canberra, 119-34.

Beer, A., Maude, A. and Pritchard, B. 2003. *Developing Australia's Regions: Theory and Practice*. UNSW Press, Sydney.

Beeton, S. 2001. Lights, camera, re-action: How does film-induced tourism affect a country town?, in *The Future of Australia's Country Towns*. Edited by Rogers, M. and Collins, Y., Centre for Sustainable Rural Communities, Latrobe University, Melbourne, 172-83.

Beeton, S. 2003. *Film-induced Tourism*. Channel View, Clevedon.

Benson, M. 2009. A desire for difference: British lifestyle migration to Southwest France, in *Lifestyle Migration: Expectations, Aspirations and Experiences*. Edited by Benson, M. and O'Reilly, K., Ashgate, Farnham, 121-35.

Berry, B. 1976. *Urbanization and Counterurbanization*. Sage, Beverly Hills.

Bolton, N. and Chalkley, B. 1990. The rural population turnaround: A case study of North Devon. *Journal of Rural Studies*, 6, 29-43.

Brennan-Horley, C., Gibson, C. and Connell, J. 2007. The Parkes Elvis Revival Festival: economic development and contested place identities in rural Australia. *Geographical Research*, 38, 73-93.

Brown, L. 2006. *Orbiting Oberon: Contemporary Migration in an Australian Country Town*. unpublished BA Honours thesis, University of Sydney.

Burnley, I. 1988. Population turnaround and the peopling of the countryside? Migration from Sydney to country districts in New South Wales. *Australian Geographer*, 19, 268-83.

Burnley, I. and Murphy, P. 1995. Residential location choice in Sydney's perimetropolitan region. *Urban Geography*, 16, 123-43.

Burnley, I. and Murphy, P. 2002. Change, continuity or cycles: The population turnaround in New South Wales. *Journal of Population Research*, 19, 137-54.

Burnley, I. and Murphy, P. 2004. *Sea Change. Movement from Metropolitan to Arcadian Australia*. UNSW Press, Sydney.

Cairns, R. 1991. *Time for a Fresh Start? Migration to Merimbula*. unpublished BA Honours thesis, University of Sydney.

Caldwell, T., Jorm, A. and Dear, K. 2004. Suicide and mental health in rural, remote and metropolitan areas in Australia. *Medical Journal of Australia*, 181 (7), S10-S14.

Cassel, S. 2008. Trying to be attractive: Image building and identity formation in small industrial municipalities in Sweden. *Place Branding and Diplomacy*, 4, 102-114.

Champion, T. 1998. Studying counterurbanisation and the rural population turnaround, in *Migration into Rural Areas: Theories and Issues*. Edited by Boyle, P. and Halfacree, K., Wiley, Chichester, 21-40.

Clout, H. 1974. The growth of second-home ownership: An example of seasonal suburbanization, in *Suburban Growth: Geographical Processes at the Edge of the Western City*. Edited by Johnson, J., Wiley, Chichester, 101-27.

Connell, J. 1972. Amenity societies: The preservation of central Surrey. *Town and Country Planning*, 405, May, 265-8.

Connell, J. 1978. *The End of Tradition: Country Life in Central Surrey*. Routledge, London.

Connell, J. and Gibson, C. 2011. Elvis in the Country: Transforming place in rural Australia, in *Festival Places: Revitalising Rural Australia*. Edited by Gibson, C. and Connell, J., Channel View, Bristol, 175-93.

Connell, J. and Rugendyke, B. 2010. Creating an authentic tourist site? The Australian Standing Stones, Glen Innes. *Australian Geographer*, 41, 87-100.

Costello, L. 2007. Going bush: The implications of urban–rural migration. *Geographical Research*, 45, 85-94.

Costello, L. 2009. Urban–rural migration: Housing availability and affordability. *Australian Geographer*, 40, 219-33.

Country Shire Councils Association and Country Urban Councils Association Working Party 1990. *Country Towns: A Future or Funeral? Report of a joint Country Shire Councils Association and Country Urban Councils Association Working Party*. Country Shire Councils Association and Country Urban Councils Association, Bunbury.

Creagh, S. and Nixon, S. 2008. The Great Sydney exodus. *Sydney Morning Herald*, 3 March, 1.

Cruickshank, J., Lysgard, H. and Magnussen, M. 2009. The logic of the construction of rural politics: Political discourses on rurality in Norway. *Geografiska Annaler B*, 91, 73-89.

Cuomo, M. 2008. *The Treechange Experience: Migration to Country New South Wales*. unpublished BA Honours thesis, University of Sydney.

Curry, G., Koczberski, G. and Selwood, J. 2001. Cashing out, cashing in: Rural change on the south coast of Western Australia. *Australian Geographer*, 32, 109-24.

Danaher, M. 2008. Seeing the change in a sea change community: Issues for environmental managers. *Australasian Journal of Environmental Management*, 15, 51-60.

Davidson, N. 2008. Selling Glen. *Glen Innes Examiner*, 7 August, 3.

Davies, A. 2008. Declining youth in-migration in rural Western Australia: The role of perceptions of rural employment and lifestyle opportunities. *Geographical Research*, 46, 162-71.

Davies, A. 2011. Local leadership and rural renewal through festival fun: the case of Snowfest, in *Festival Places: Revitalising Rural Australia*. Edited by Gibson, C. and Connell, J., Channel View, Bristol, 61-73.

Davison, G. 2005. Rural sustainability in historical perspective, in *Sustainability and Change in Rural Australia*. Edited by Cocklin, C. and Dibden, J., UNSW Press, Sydney, 38-55.

Dickins, J. 2007. People move for the water: New trend of 'oasis change'. *The Sunday Telegraph*, 17 June, 37.

Digby, B. 2004. Beyond the farm gate: Changing rural economies and lifestyles in Australia. *GeoDate*, 17(3), 5-8.

Dollery, B., Crase, L. and O'Keefe, S. 2009. Improving efficiency in Australian local government: Structural reform as a catalyst for effective reform. *Geographical Research*, 47, 269-79.

Drew, P. 1994. *The Coast Dwellers. Australians Living on the Edge*. Penguin, Melbourne.

Drysdale, R. 1991. Aged migration to coastal and inland centres in NSW. *Australian Geographical Studies*, 29, 268-84.

Duggan, A. 2005. Want a treechange? *Sunday Mail Magazine*, 19 June, 43.

Easthope, H. and Gabriel, M. 2008. Turbulent lives: Exploring the cultural meaning of regional youth migration. *Geographical Research*, 46, 172-82.

Escribano, M. 2007. Migration to rural Navarre: Questioning the experience of counterurbanisation. *Tijdschrift voor Economische en Sociale Geografie*, 98, 32-41.

Essex, S. and Brown, G. 1997. The emergence of post-suburban landscapes on the north coast of New South Wales: A case study of contested space. *International Journal of Urban and Regional Research*, 21, 259-85.

Fisher, T. 2003. Differentiation of growth processes in the Peri-urban region: An Australian case study. *Urban Studies*, 40, 551-65.

Fitzgerald, R., Megarrity, L. and Symons, D. 2009. *Made in Queensland: A New History*. University of Queensland Press, Brisbane.

Forth, G. and Howell K. 2002. Don't cry for me Upper Wombat: The realities of regional/small town decline in non coastal Australia, *Sustaining Regions*, 2, 4-11.

Fountain, J. and Hall, M. 2002. The impact of lifestyle migration on rural communities: A case study of Akaroa, New Zealand, in *Tourism and Migration*. Edited by Hall, C. and Williams, A., Kluwer, Amsterdam,153-68.

Freestone, R. 1989. *Model Communities: The Garden City Movement in Australia*. Nelson, Sydney.

Gabriel, M. 2002. Australia's regional youth exodus. *Journal of Rural Studies*, 18, 209-12.

Ghose, R. 2004. Big sky or big sprawl? Rural gentrification and the changing cultural landscape of Missoula, Montana. *Urban Geography*, 25, 528-49.

Gibson, C. and Argent, N. 2008. Getting on, getting up and getting out? Broadening perspectives on rural youth migration. *Geographical Research*, 46, 135-8.

Gibson, C. and Connell, J. 2003. 'Bongo fury': Tourism, music and cultural economy at Byron Bay, Australia. *Tijdschrift voor Economische en Sociale Geografie*, 94, 164-87.

Gibson, C. and Connell, J. 2011. *Music Festivals and Regional Development in Australia*. Ashgate, Farnham (in press).

Grigg, A. 2006. Endangered species: Australia's disappearing country towns. *Australian Financial Review*, 4 November, 19-21.

Gurran, N. 2008. The turning tide: Amenity migration in coastal Australia. *International Planning Studies*, 13, 391-414.

Gurran, N. and Blakely, E. 2007. Suffer a sea change? Contrasting perspectives towards urban policy and migration in coastal Australia. *Australian Geographer*, 38, 113-31.

Halfacree, K. 1993. Locality and social representation: space, discourse and alternative definitions of the rural. *Journal of Rural Studies*, 9, 23-37.

Halfacree, K. 1997. Contrasting roles for the post-productivist countryside, in *Contested Countryside Culture*. Edited by Cloke, P. and Little, J., Routledge, London, 70-93.

Halfacree, K. 2004. A utopian imagination in migration's *Terra Incognita*? Acknowledging the non-economic worlds of migration decision-making. *Population, Space and Place*, 10, 239-53.

Hall, L. 2006. Tug-of-war for a battling town's youngest adults. *The Sun-Herald*, 23 July, 16-17.

Hamilton, C. and Mail, E. 2003. Downshifting in Australia: A sea change in the pursuit of happiness, Australia Institute Discussion Paper No. 50, Canberra.

Harley, R. and Phillips, M. 2006. If you go down to the woods today ... *Australian Financial Review*, 13 April, 20.

Healy, K. and Hillman, W. 2008. Young families migrating to non-metropolitan areas. *Australian Journal of Social Issues*, 43, 479-97.

Higgins, E. 2009. Ring the bell for Nimmity. *The Australian*, 30 December, 9.

Hillman, K. and Rothman, S. 2007. Movement of Non-metropolitan Youth towards the Cities, ACER Research Report No. 50, Australian Council for Educational Research, Melbourne.

Hjort, S. and Malmberg, G. 2006. The attraction of the rural: Characteristics of rural migrants in Sweden. *Scottish Geographical Magazine*, 122, 55-75.

Hoey, B. 2009. Pursuing the good life: American narratives of travel and a search for refuge, in *Lifestyle Migration: Expectations, Aspirations and Experiences*. Edited by Benson, M. and O'Reilly, K., Ashgate, Farnham, 31-50.

Holloway, S. 2007. Burning issues: whiteness, rurality and the politics of difference. *Geoforum*, 38, 7-20.

Hughes, A. 2008. *Art life Chooks: Learning to leave the city and love the country*. HarperCollins, Sydney.

Hugo, G. 1994. The turnaround in Australia: Some first observations from the 1991 census. *Australian Geographer*, 25, 1-17.

Hugo, G. 2005. The state of rural populations. in *Sustainability and Change in Rural Australia*. Edited by Cocklin, C. and Dibden, J., UNSW Press, Sydney, 56-79.

Hugo, G. and Bell, M. 1998. The hypothesis of welfare-led migration to rural areas, in *Migration into Rural Areas*. Edited by Boyle, P. and Halfacree, K., Wiley, New York, 107-33.

James, K. 1989. Work, Life and Leisure, in *Women in Rural Australia*. Edited by K. James, University of Queensland Press, Brisbane, 67-81.

Jeans, D. 1972, *An Historical Geography of New South Wales to 1901*. Reed Education, Sydney.

Jensen, E. 2009. From boom town to bust. *Sydney Morning Herald*, 24 January, 4.

Jetzkowotz, J., Schneider, J. and Brunzel, S. 2007. Suburbanisation, mobility and the 'good life in the country': A lifestyle approach to the sociology of urban sprawl in Germany. *Sociologia Ruralis*, 47, 148-71.

Jones, G. 1997. Greg Jones' Moree, in *Heart of the Country*. Edited by I. Hamilton, Wakefield Press, Adelaide.

Jopson, D. 2009. Raining hope out west. *Sydney Morning Herald*, 5 December, 9.

Kapferer, J. 1990. Rural myths and urban ideologies. *Australian and New Zealand Journal of Sociology*, 26, 87-107.

Karn, V. 1977. *Retiring to the Seaside*. Routledge and Kegan Paul, London.

Kempsey Shire Council 2009. Mayoral report: Country and regional living expo 7th-9th August, 2009. File 233 Mayor. Available at http://www.kempsey.nsw. gov.au/pdfsCnlMtgs/2009/ (accessed 10 December 2009).

Kijas, J. 2002. A place at the coast: Internal migration and the shift to the coastal-countryside. *Transformations*, 2, March, 1-12.

Lewis, D. 2005. Greener pastures. *Sydney Morning Herald*, 1 October, 21, 27-8.

Lewis, G. 1998. Rural migration and demographic change, in *The Geography of Rural Change*. Edited by Ilbery, B., Longman, Harlow, 131-60.

Lindgren, U. 2003. Who is the counter-urban mover? Evidence from the Swedish urban system. *International Journal of Population Geography*, 9, 399-418.

Lockie, S. 2000. Crisis and conflict: shifting discourses of rural and regional Australia, in *Land of Discontent*. Edited by Pritchard, B. and McManus, P., UNSW Press, Sydney, 14-32.

Lowe, P. and Ward, N. 2009. England's rural futures: A socio-geographical approach to scenarios analysis. *Regional Studies*, 43, 1319-32.

Lundholm, E., Garvill, J., Malmberg, G. and Westin, K. 2004. Forced or free movers? The motives, voluntariness and selectivity of migration in the Nordic countries. *Population, Space and Place*, 10, 59-72.

McManus, P. 2005. *Vortex Cities to Sustainable Cities: Australia's Urban Challenge*. UNSW Press, Sydney.

McManus, P. and Pritchard, B. 2000. Geography and the emergence of rural and regional Australia. *Australian Geographer*, 31, 383-91.

Maginn, P. and Rofe, M. 2007. Urbanism and regionalism down under: An introduction. *Space and Polity*, 11, 201-8.

Maher, C. and Stimson, R. 1994. *Regional Population Growth in Australia: Nature, Impact and Implications*. Australian Government Publishing Service, Canberra.

Marsden, B. 1969. Holiday homescapes of Queensland. *Australian Geographical Studies*, 7, 57-73.

Maude, A. 2003. Local and regional economic development organisations in Australia, in *Developing Locally: An International Comparison of Local and Regional Economic Development*. Edited by Beer, A., Haughton, G. and Maude, A., Policy Press, Bristol, 109-36.

Mayes, F. 1996. *Under the Tuscan Sun*. Broadway Books, New York.

Mayes, R. 2008. A place in the sun: The politics of place, identity and branding. *Place Branding and Public Diplomacy*, 4, 124-35.

Mayle, P. 1990. *A Year in Provence*. Knopf, New York.

Milbourne, P. (ed.) 1997. *Revealing Rural Others: Representation, Power and Identity in the British Countryside*. Pinter, London.

Milbourne, P. 2007. Re-populating rural studies: Migrations, movements and mobilities. *Journal of Rural Studies*, 23, 381-6.

Ministry of Agriculture and Forestry (MAF) 1994. *New Zealand Regional Diversity*. MAF, Wellington.

Mitchell, C. 2004. Making sense of counterurbanization. *Journal of Rural Studies* 20, 15-34.

Montague, M. 1980. Barcaldine: A Queensland rural community, in *Mobility and Community Change in Australia*. Edited by Burnley, I., Pryor, R. and Rowland, D., University of Queensland Press, Brisbane, 19-37.

Moree Plains Shire Council 2009. Ordinary Council Meeting Minutes. Reports: 2. *Healing Waters Express Report*, 8-14.

Morris, L. 1999. Flight of the battlers. *Sydney Morning Herald*, 22 May, 1, 7.

Munro, P. 2009. When the tree-change dream turns to dust. *The Age*, 19 April.

Murphy, P. 2006. Seachange to hillchange: A new equilibrium?, in *The Changing Natures of Australia's Country Towns*. Edited by Rogers, M. and Jones, D., Victorian Universities Regional Research Network Press, Ballarat, 26-37.

Narromine Shire Council 2007. Narromine tree changers, Narromine, ms.

Neal, S. 2009. *Rural Identities: Ethnicity and Community in the Contemporary English Countryside*. Ashgate, Farnham.

New South Wales Government 1995. *New South Wales Social Trends Bulletin. Issue 2: Rural Social Trends*. New South Wales Government, Sydney.

Newby, H. 1979. *Green and Pleasant Land? Social Change in Rural England*. Hutchinson, London.

Newell, P. 2000. *The Olive Grove*. Penguin, Melbourne.

Niedomysl, T. 2004. Evaluating the effects of place marketing campaigns on inter-regional migration in Sweden. *Environment and Planning A*, 36 1991-2009.

Niedomysl, T. 2007. Promoting rural municipalities to attract new residents: An evaluation of the effects. *Geoforum*, 38, 698-709.

Ni Laoire, C. 2007. The 'green green grass of home'? Return migration to rural Ireland. *Journal of Rural Studies*, 23, 332-44.

Oberon Information Centre 2009. *Oberon ... Simply Spectacular*. Oberon Information Centre, Oberon.

O'Connor, K. 2001. Coastal development: Just a little shift in Australia's geography. *People and Place*, 9, 49-56.

O'Dwyer, E. 2008. Surely there can't be hash in that quiche? *The Sun-Herald*, 17 August, 8.

Osti, G. 2010. Rural melting pots, mobilities and fragilities: Reflections on the Spanish case. *Sociologia Ruralis*, 50, 277-95.

Pahl, R. 1965. *Urbs in Rure. The Metropolitan Fringe in Hertfordshire*. LSE Geographical Papers No. 2, London.

Panelli, R. 2001. Narratives of community and change in a contemporary rural setting: The case of Duaringa. *Australian Geographical Studies*, 39, 156-66.

Paquette, S. and Domon, G. 2004. Changing ruralities, changing landscapes: exploring social recomposition using a multi-scale approach. *Journal of Rural Studies*, 19, 425-44.

affluence of *Country Style* (or even of *Live the Dream*) despite several councils bringing their wines and gourmet foods to the Expos. Gentrification is certainly not unwelcome – professions are equally in demand – but it has not been the priority of many councils.

The notion of moving to the country, and living the dream, is founded on particular visions of the country and a certain rural ambience. The images used to promote CW have a wholly middle class and lifestyle focus. Such bucolic images of rural life are potentially at odds with the structure of migration to regional areas, the actual work performed by people who respond to the advertisements on the Jobs Board (fitters and turners, accountants, plumbers, nurses, receptionists), and with the image of a rural and regional Australia that has experienced problems.

Harsh Realities and Tough Geographies

While CW naturally projects a positive image of rural life, rather than focusing on challenges or hardship, government policies have rarely favoured people living in rural settlements as social services have been withdrawn, centralised or privatised. Global shifts in the terms of trade have disadvantaged some agricultural producers, but rural is not necessarily equated with agricultural and various parts of rural and regional Australia have prospered on the back of a mining resources boom. Unusually for a high income country, regional Australia remains invaluable to the health of the nation, rather than a place of mere historical and sentimental interest. Half of Australia's export wealth comes from commodities mainly grown or extracted in rural and regional areas while two thirds of the nation's continental water supplies are managed by farmers. Agriculture may generate little employment but it is vital to the economic and social health of many rural towns.

Yet virtually throughout predominantly agricultural areas, significantly in the great wheat and sheep belts, there has been a shift towards a post-productivist countryside. Not only has that shift been unwelcome to traditionalists but, as in the small Queensland town of Duaringa, the decline of agriculture led to services becoming the principle rationale for economic survival, and a sense that the real meaning of the place as a centre of production had simultaneously been lost (Panelli 2001). Despite *Live the Dream*'s editorial, countrymindedness as an idea that celebrated the life and work of rural producers has also largely disappeared. Regional life is consequently more problematic than in earlier decades when 'agriculture promised to return civilisation to the frontier' and appeared 'more suited to a modern state than squatting or mining' (Muir 2010: 63), agricultural futures seemed secure, political support was present for the places and export industries that characterised Australia and rural people were committed to similar priorities and to their local communities.

Australia has become a thoroughly urban nation, almost without global precedent, clinging to the narrow coastal veranda of a vast continent. Perhaps, in this urban world, 'we should not forget the critical role that rural Australia has

Phillips, M. 2002. The production, symbolization and socialization of gentrification: A case study of a Berkshire village. *Transactions of the Institute of British Geographers*, 27, 282-308.

Phillips, M. 2004. Other geographies of gentrification. *Progress in Human Geography*, 28, 5-30.

Pritchard, B. and McManus, P. (eds) 2000. *Land of Discontent: The Dynamics of Change in Rural and Regional Australia.* UNSW Press, Sydney.

Pryor, L. and Lewis, D. 2004. Sydney it ain't and that's exactly why they love it, *Sydney Morning Herald*, 14 August, 25.

Queensland Government 2010. Estimated Population by Urban Centre and Locality: Queensland, 2001 to 2009. (Available at www.oesv.qld.gov.au/products/tables).

Rawsthorne, M., Hillman, W. and Healy, K. 2009. Families on the fringe: Mental health implications of the movement of young families to non-metropolitan areas. *Rural Society*, 19, 306-17.

Roberts, R. 1995. A 'fair go for all'? Discrimination and the experiences of some men who have sex with men in the bush, in *Communication and Culture in Rural Areas.* Edited by Share, P. CSU Centre for Rural Social Research, Wagga Wagga, 151-74.

Rockhampton Regional Development Limited 2007. *Central QueensLand of Opportunity: Twelve Lifestyle Adventurers Living The Dream.* Rockhampton RDL, Rockhampton.

Salt, B. 2001. *The Big Shift.* Hardie Grant, Melbourne.

Salt, B. 2006. *The Big Picture.* Hardie Grant, Melbourne.

Salt, B. 2009. Boomers pound the coast. *The Australian*, 10 October, 2.

Sant, M. and Simons, P. 1993. Counterurbanization and coastal development in New South Wales. *Geoforum*, 24, 291-306.

Share, P. 1995. Beyond 'countrymindedness': Representation in the post-rural era, in *Communication and Culture in Rural Areas.* Edited by Share, P. CSU Centre for Rural Social Research, Wagga Wagga, 1-23.

Short, J. 1991. *Imagined Country.* Routledge, London.

Smailes, P. 1997. Socio-economic change and rural morale in South Australia, 1982-1993. *Journal of Rural Studies*, 13, 19-42.

Smailes, P. 2002. From rural dilution to a multifunctional countryside: Some pointers to the future from South Australia. *Australian Geographer*, 33, 79-95.

Smith, D. 2007. The changing faces of rural populations: '"(re)Fixing" the gaze' or 'eyes wide shut'? *Journal of Rural Studies*, 23, 275-82.

Smith, J. 2004. *Australia's Rural and Remote Health: A Social Justice Perspective.* Tertiary Press, Melbourne.

South Burnett Regional Council 2008. The Past, The Present and the Future, Murgon, ms.

Statistics New Zealand 2008. *New Zealand: An Urban/Rural Profile Update 2001-2008.* Statistics New Zealand, Wellington.

Stockdale, A. 2010. The diverse geographies of rural gentrification in Scotland. *Journal of Rural Studies*, 26, 31-40.

Storey, D. 2004. A sense of place: Rural development, tourism and place promotion in the Republic of Ireland, in *Geographies of Rural Cultures and Societies*. Edited by Holloway, L. and Kneafsey, M., Ashgate, Aldershot, 197-213.

Strathern, M. 1984. The social meaning of localism, in *Locality and Rurality: Economy and Society in Rural Regions*. Edited by Bradley, T. and Lowe, P. Geo Books, Norwich, 181-97.

Strong, K., Trickett, P., Titulaer, I. and Bhatia, K. 1998. *Health in Rural and Remote Australia*. Australian Institute of Health and Welfare, Canberra.

Swaffield, S. and Fairweather, J. 1998. In search of Arcadia: The persistence of the rural idyll in New Zealand rural subdivisions. *Journal of Environmental Planning and Management*, 41, 111-27.

Taskforce on Regional Development 1993. *Developing Australia: A Regional Perspective – A report to the Federal Government by the Taskforce on Regional Development Volumes 1 and 2*. Taskforce on Regional Development, Canberra.

Thomas, D. 1970. *London's Green Belt*. Faber and Faber, London.

Tonts, M. 2005. Competitive Sport and Social Capital in Rural Australia. *Journal of Rural Studies*, 21, 137-49.

Tsioutis, A. 2007. *Choosing to Live the Dream? An Analysis of Country Week*. unpublished BSc Honours thesis, University of Sydney.

Turner, G. 2008. Editor's letter, *Live the Dream: Sea and Tree Change Australia*, 1, 2.

Vergunst, P. 2009. Whose Socialisation? Exploring the Social Interaction between Migrants and Communities of Place in Rural Areas. *Population, Space and Place*, 15, 253-66.

Victorian Government 2007. *Make it Happen in Provincial Victoria*. http://www.provincialvictoria.vic.gov.au/about.aspx (accessed 27 November 2009).

Walmsley, D.J, Epps, W. and Duncan, C. 1998. Migration to the New South Wales North Coast 1986-1991: Lifestyle motivated counterurbanisation. *Geoforum*, 29, 105-18.

Waterhouse, R. 2002. Rural culture and Australian history: Myths and realities, *Arts*, 24, 83-102.

Wendt, S. 2009. *Domestic Violence in Rural Australia*, Federation Press, Sydney.

Wild, R. 1974. *Bradstow: A Study of Status, Class and Power in a Small Australian Town*. Angus & Robertson, Sydney.

Wilkinson, M. and Cubby, B. 2009. Water crisis in the west as Lachlan River runs dry, *Sydney Morning Herald*, 24 October, 3.

Witherby, A. 2001. Supermarkets – Scourge or saviour?, in *The Future of Australia's Country Towns*. Edited by Rogers, M. and Collins,Y. Latrobe University Centre for Sustainable Rural Communities, Melbourne, 193-207.

Wood, R. 2008. *Survival of Rural America: Small Victories and Bitter Harvests*. University Press of Kansas, Lawrence.

Zukin, S. 1995. *The Cultures of Cities*. Blackwell, Oxford.

Index

abattoir 5, 26, 79, 82, 115
Aborigines 3, 10, 15, 83, 113, 118, 120, 163, 178
acreage 117, 123, 125, 127, 131
advertising xiv, 21, 33, 39-40, 55, 59, 86, 106, 138, 143, 170-71
aesthetics 32, 41, 130
affluence 24-5, 26-7
affordability 107, 108, 109-10, 116, 120, 132, 140, 151, 163
aging 5, 10, 13, 16-17, 29, 56, 58, 117, 125, 175
agrarianism, *see also* countrymindedness 5, 19, 170
agriculture 1-5, 10, 13-16, 25, 32, 34, 43-4, 52, 57, 94, 123, 149, 152, 163-5, 171-2
Airlie Beach, Queensland (Q) 26
airports 102, 103, 118, 120
Albury, NSW (including Albury-Wodonga) 16, 23, 54, 100
alcohol 15
amenity 19, 27, 34-5, 174, 176
anonymity 29, 128
anti-urbanism 28, 42, 77-8, 98-9, 111-12, 172, 176
Australian Celtic Festival, *see also* Celtic Country 66, 90, 127-8
Ariah Park, NSW 14, 172-3
Armidale, NSW (including Armidale-Dumaresq) 8, 16, 40, 45, 51, 54, 61, 89, 99, 100, 103, 104, 105, 106, 110, 118, 120, 122, 132, 139-40, 142-3, 144, 149, 150, 151-2, 153, 154
Australian Gourmet Traveller 32
Australian Labour Party (ALP) 5, 21, 41

Bailey, Peter 40, 42, 59, 72-3, 74, 77, 97, 139, 154
Ballina, NSW 27, 28, 100

banks 88, 118, 127
Barcaldine, Q 8, 9, 134
Barossa Valley, SA 32
Barwon Heads, Victoria 30-31
Bathurst, NSW 8-9, 15, 20, 24, 34, 61, 93, 100, 101, 102, 118, 120, 122, 124, 126, 131-2, 153, 172
battlers 43
beaches 43, 103
belonging 14, 130, 133-5
Berry, NSW 131-2
Blackall, Q 79
Blue Mountains, NSW 24, 100, 122, 155
Boorowa, NSW 46, 100, 157
boredom 128, 167
Bourke, NSW 59, 164
Brisbane 2-4, 12, 15, 26, 41-4, 50, 53, 101, 111, 117, 120, 122, 129, 172, 174
Britain, *see also* England, Scotland and Wales 13, 23, 24, 28, 36, 120, 133, 135
Broken Hill, NSW 4, 8-9, 20, 51
Broken Hill Proprietary (BHP) 52
'bush, the' 1-2, 4, 14, 21, 32, 34, 36, 42-3, 173, 177
bushfires 23, 36, 163
business 6, 10, 45, 54-6, 59, 61, 86, 88, 90-92, 94-5, 108, 117-18, 120, 123, 126, 131, 139, 142-3, 149-50, 156, 159, 162, 163, 165, 169, 176-7
Byron Bay, NSW 8, 26-8, 31

Cabonne, NSW 46
C-Change 46
Cairns, Q 8, 9, 27, 102, 104, 157
Canada 1-2, 12, 134
Canberra, ACT 6, 15, 24, 49, 51, 61, 100, 105, 122, 172
capitalism 21, 36
cappuccino, *see* coffee
Capricorn Coast, Q 27

Castlemaine, Victoria 130, 133
caravans 27
cars 24, 118, 124, 132
Celtic Country 48, 118, 133
Charleville, Q 9, 12
cheese 5, 32
children 15-17, 36, 41, 49, 52, 58-9, 87,
        89, 98, 105, 109, 110, 111-12, 114,
        116, 119, 122-3, 125, 127, 129-32,
        139-40, 141-2, 144, 146, 148, 158,
        175
Chinchilla, Q 54
churches 17, 144
city 4, 12, 18-22, 41, 43-4, 55, 59, 87-8,
        95, 98, 99, 124, 128, 130, 132, 134,
        141, 143, 154-5, 159-60, 169-70,
        172-5, 177-8
class, *see* social class
climate 5, 28, 29, 48, 121-4, 127, 151, 155
climate change 2, 164
coaches 157-8
coasts xiv, 1, 6, 8-9, 12-13, 20, 22, 24-5,
        28, 39, 43-4, 48, 58-61, 101, 103,
        115, 117, 122-4, 132, 135, 162,
        171-5, 177-8
Cobar, NSW 10
coffee 43, 77, 82, 87, 128, 153, 161, 174
Coffs Harbour, NSW 8, 26, 27, 48, 51, 58,
        100, 157
colonialism 3-4
communes 25
community 10, 15-17, 23, 25, 29, 31, 35,
        41-3, 46, 50, 54-6, 58-9, 85-9, 92,
        94-5, 99, 105-6, 123-32, 133-5,
        139-43, 144-6, 155-6, 158, 167,
        176
commuting 1, 5-6, 9, 19, 24, 28, 39, 42-3,
        48, 61, 107, 128, 141, 144, 156,
        170
Condobolin, NSW 8, 11, 100, 158, 164
congestion 22, 27, 35, 41-2, 44, 58, 101,
        111, 124, 128, 177
conservation 27, 110, 130, 164
conservatism 53, 134-5, 161, 176, 178
consumption 25, 28, 130, 132, 174
Cooma, NSW 6, 51, 72, 75-6, 83, 100
Coonamble, NSW 61

Cootamundra, NSW 8, 46, 50-51, 65, 69,
        78, 83, 89, 100, 141, 143-5, 149,
        151, 153, 155, 173
councils, *see also* government, local 1-2,
        10, 16-17, 21, 39-61, 63-9, 73-96,
        171-4, 176
counter-urbanisation xiv, 9, 12, 18, 26-30,
        39, 43-4, 95, 117, 169-70, 177
counter-urbanism 27, 28
'country, the' 2-5, 18-23, 31-7, 40-44, 55,
        57, 59, 88, 95, 122-5, 127-30, 132-
        3, 167-75, 178
Country Party, *see also* National Party 4,
        18-20, 39
Country Style 21, 32, 157, 171
Country Week, *see also* Expo xiv, 2, 10,
        16, 21, 39-62, 85-96, 117-18, 125,
        137-44, 164-5, 170-77
Country Women's Association (CWA) 52,
        161, 174
countrymindedness xv, 2, 18-20, 23, 41-2,
        167, 171-3
Cowra, NSW 103, 109, 113, 164
cricket 17, 23, 46, 50, 118
crime 98, 99, 111, 124, 128, 132, 141, 144,
        155, 164, 176
Crookwell, NSW 100, 143-4
cultural capital 29
culture 14, 26-9, 36, 59, 90, 132, 153, 174
Cumnock, NSW 53-4, 134, 173
Cunnamulla, Q 9, 146

Dalby, Q 9, 50, 54, 89, 103-4, 106, 108,
        164
Daysdale, NSW 17
decentralisation 3, 5, 22, 23, 24, 156, 167,
        173
'deep rural' 19, 172
demography, *see* population
Deniliquin, NSW 8, 59
Denmark, WA 36, 148
dentists 76, 132, 150, 153
Depression, The Great 3-4
distance 3, 5-6, 39, 59, 100, 101-4, 115,
        120, 122, 124, 125, 127, 132, 139,
        141, 153, 172, 174-5
divorce 119, 125, 164

doctors 151, 159, 164

downshifting 27-8, 124, 172

drought 2-4, 10, 14, 19, 42, 72-3, 88, 107, 127, 162-4, 167, 177

drugs 155

Duaringa, Q 171

Dubbo, NSW 6, 34, 46, 51-2, 54, 73, 78, 83, 100, 103, 122, 148, 153-4

DURD (Department of Urban and Regional Development) 23

economy 3, 5, 10, 16-7, 20-22, 39-42, 50, 52-3, 55, 59, 75-6, 85, 90, 92-4, 117, 121, 125, 135, 169, 171-8
   economic decline 3, 13, 19, 21, 41, 53, 162-4, 169
   economic growth 6, 10, 20-21, 41, 52, 55-6, 59, 90, 130, 149-50, 178

education 13-6, 20-21, 43, 51, 53, 87-8, 109, 110, 121, 132, 140, 144, 148, 153, 158, 177-8

electricity 78, 84, 123, 127, 177

elites 32-4, 36, 53, 72, 131

Emerald, Q 9, 56

emotion 102, 133

employment xiii, 1-2, 4-5, 10, 12-16, 20, 28, 29, 39, 41-4, 46, 52-9, 86-7, 90-92, 96, 98, 101, 102, 108-9, 115, 116, 117, 119-22, 124-6, 128, 135, 140, 149, 150-51, 170-71, 173-6, 178

England 3, 23-4, 28, 133, 174

entertainment 97, 135, 165

environment 3, 18-19, 27, 29, 58, 98, 107, 121, 123, 127, 130, 132, 140, 144, 152, 159, 162, 173-4

environmental degradation 27

environmental determinism 19

equality 20

escape 14, 59, 74

estate agents 29, 50, 52, 54, 85-6, 88, 90, 92-3, 125

ethnicity 26, 112-13, 120, 147-8, 178

Expo xiii, 2, 10, 21, 39-42, 44-8, 50-118, 135, 170-78

families xiii, 14-16, 28, 36, 41-3, 46, 52-3, 55-7, 59, 87, 89, 93-4, 96, 98, 102-

3, 105, 110, 119-20, 122-6, 128, 130-33, 137, 139-43, 146-8, 150, 158, 166, 169-70, 172-8

farmhouses 28, 32, *see also* rentafarmhouse

farms and farming 3, 5, 9-10, 17, 19-20, 33, 123, 127, 130, 171-3, 175

festivals 25, 46, 50, 66, 69, 89-90, 94-5, 101, 127-8, 134, 141, 157, 176

fishing 26, 131, 152, 158, 161

flies 161

flood 23

food production 5, 141
   consumption 3-5, 18, 25-6, 35-6, 54, 59, 61, 66, 77, 123, 132, 153, 157, 159, 161, 171, 176
   purchase 132, 153, 161

football 17, 118, 131, 146, 158, 159

Forbes, NSW 8, 11, 51, 100, 164

Forster, NSW 100

Foundation for Regional Development, *see* Country Week

France 25, 33, 167

freedom 74, 127, 132

friends 15, 98, 110, 120, 122-3, 126, 129, 135, 144-6, 162, 165, 175

gardens 68, 122, 127, 131-2, 152, 175

gentrification xiii, xiv, 18, 24, 170-71

Gilgandra, NSW 100

girls 13

Glen Innes 8, 11, 14, 40, 46-7, 49, 51, 65-9, 90, 100, 117-32, 135, 149, 153, 169, 175

Gloucester, NSW 100

Gold Coast, Q 8, 26, 27

golf 59, 118, 131, 145, 159, 161

Goulburn, NSW 4, 66, 139, 154

government 6, 13, 17, 39-41, 53, 94, 163, 171, 176
   Commonwealth/Federal/National 1, 5, 39, 41
   local 1-2, 10-11, 14, 41, 45, 50, 53, 55, 95, 117-18, 177
   state 1, 5, 39-41, 44, 51-2, 54, 82, 95

Grenfell, NSW 89-90, 94-5, 100, 105, 106, 144, 145, 146

Griffith, NSW 148

growth centres 6, 18, 23-4
Gunnedah, NSW 4, 8, 10, 14, 48-9, 51, 56, 94, 100, 151
Guyra, NSW 152
Gwydir, NSW 51, 61, 100, 106, *see also* Warialda

Hanson, Pauline 2
health 69, 71, 87, 105, 107, 110, 119, 122, 128, 132, 141, 143, 146, 147, 148, 153, 159, 162, 164, 171
    employment and workers 49, 51, 53, 76, 110, 121-2, 162
    facilities and services 16, 21, 63, 68, 72, 79, 105, 132, 153
heritage 35, 59, 89, 152, 157, 176
hippies 148, 160
hobby farms 123, 127
homes 1, 10, 14, 16, 24, 42, 46, 49, 61, 87, 118, 125-7, 133-4, 174-5, 178
homosexuality (includes gay, lesbian) 36, 147-8, 166, 178
horses 74, 103, 132, 152, 159, 161
hospitals 49, 51, 102, 110, 118, 144, 153, 173
housing 41-3, 46, 52, 77, 87, 92-3, 98, 101, 109-10, 116, 117, 124-6, 135, 170, 174-6, 178
    price 21, 24, 25, 27, 31, 35, 43, 50, 64, 89, 98, 109, 118, 119, 121-2, 126, 130, 132, 144, 151, 159-60, 170, 173
    rental 24, 64
Howard, Ebenezer 23
Howard, John (politician) 1, 43
Howard, John (actor) 31
Hughenden, Q 9, 145, 156, 165-7
Hunter Valley, NSW 4, 25, 32, 33, 51, 59, 61, 100, 174
hunting 131, 161
hybrid country 69, 72, 116, 170, 174, 178, *see also* micropolitan, urban country

identity xiv, 13-14, 17, 44, 54, 57, 84-5, 95, 130, 167, 177
ideology xiv, 20, 156, 167-8

immigration 3, 20, 37, 108-9, 112, 117, 118, 120, 176
industry 4, 10, 27, 57, 59, 71, 86, 92, 94, 118, 121, 126, 128
inertia 114, 134, 148
internet 77, 105, 115, 125, 142, 143, 154
irrigation 5, 163-4
income 17, 27, 90, 92-3, 99, 126, 163, 171
Innisfail, Q 48, 65, 73, 76, 104
Inverell, NSW 48, 51, 69, 81, 90, 100, 102-3, 118, 163
Italy 25, 120

Jesenik (Czech Republic) 46
Jobs Board 75, 170-71
Julia Creek, Q 46, 69, 77
Junee, NSW 77, 82, 174

Kangaroo Valley, NSW 25, 61, 143, 154
Kempsey, NSW 51, 54, 59
Kentucky Fried Chicken (KFC) 6, 66, 149
Kerr, Miranda 68
Kiama, NSW 26
kinship, *see* families
koalas 68, 152

Lachlan Shire 11, 51, 61, 69, 83-4, 100
Lake Macquarie, NSW 27
language 118-20
leisure 25, 59, 72, 87, 170
Liberal Party 1, 21, 23, 43
libraries 153, 156
lifestyle xiv, 13-6, 18, 21, 25-6, 27-9, 31, 36, 41, 43-4, 46, 52, 55-6, 59, 87, 91, 98, 101, 103, 106-7, 121, 122, 124-6, 128, 135, 144, 150, 157-62, 166, 169-78
Lightning Ridge, NSW 109
Limestone Coast, SA 33
Lismore, NSW 50
Lithgow, NSW 120
Liverpool, Sydney 139
Liverpool Plains, NSW 16, 48, 51, 83, 100
Live the Dream 21, 29, 32, 43, 72, 137-8, 156-65, 170-71, 177-8
'locals' 113, 118-19, 129-31, 133-4, 158

Longreach, Q 9, 12
loneliness 161
love 14, 123, 134, *see also* romance

McDonald's 6, 66, 149, 174
Mackay, Q 49
managers 15, 55, 98, 108, 150, 167
Manning Valley, NSW 49, 70, 100
manufacturing 25, 59, 94, 139, *see also* industry
markets 127-8, 132
marriage 15, 30, 119, 123, 167
Marulan, NSW 139
mateship 19
Mayes, Frances 25-6, 31-2
Mayle, Peter 25-6, 31-2
media 16, 18, 21, 25, 26, 30-31, 32-3, 36, 58, 64, 70-72, 86-8, 117, 137-62, 170, 177, *see also* newspapers, television
Melbourne, Victoria 159
Merimbula, NSW 29
micropolitan 153, 178, *see also* hybrid country, urban country
migration 2, 6, 9-10, 12, 14-16, 18, 22, 24-37, 39, 41, 43-4, 52, 61, 88-9, 92-3, 98, 117-27, 129-30, 134-5, 154, 169-78
    rationale 119, 121-5
    return migration 15-16, 134, 142, 143, 148, 155, 167, 169
    rural-urban 4-5, 20, 162
    urban-rural 2, 3, 9, 24-37, 39, 44, 118-20, 169-70, 172, 178
mining xiv, 3-5, 9-10, 12, 34, 48, 52, 55-6, 68, 94, 99, 104, 108-9, 118, 149, 165, 171
mobility 2, 24, 44, 85, 93, 135, 169
Monarto, SA 24
morality 167
Moranbah, Q 9
Moree, NSW (including Moree Plains Shire) 8, 11, 46, 48, 51, 61, 89, 100-101, 163, 173-4
Mount Isa, Q 12, 56, 84
Mudgee, NSW 8, 34, 36, 66, 100-102
Murrumbidgee River 5, 6
Murray River 5

Muswellbrook, NSW 48, 50-51, 66, 70-71, 81, 89, 99-100, 149, 174

Nambucca, NSW 59
Narrabri, NSW 8, 48, 51, 69, 78-9, 100, 149, 173
Narromine, NSW 78, 100, 137, 145, 147-8, 150-51, 153-4, 160, 173, 176
national parks 59, 71, 74, 121, 152
National Party 18, *see also* Country Party
nature 44, 59, 71, 113, 133, 147, 177
newcomers 28, 118-20, 123-35, 136-62, 169-70, 175-7
neighbours 111, 127, 129-32
Netherlands 135
Newcastle, NSW 1, 4, 16, 20, 59, 73, 100, 153-4, 174
Newell, Patrice 33
New England 40-41, 51, 66, 82, 118, 125, 147-8, 150, 154
New Zealand 1-2, 13, 36, 120, 133
newspapers 16, 33, 42-3, 50-51, 54, 59, 64, 70-72, 85-86, 135, 137-56, 162, 164, 170, *see also* media
Nimbin, NSW 25-7, 134, 148
Nimmitabel, NSW 6, 172
noise 77, 127-8, 151, 160
Noosa, Q 26
Northern Rivers, NSW 100
Norway 167
nostalgia 21, 31, 35, 77-8, 111, 116, 167, 172, 176
Nundle, NSW 82, 162
nurses 162, 164

Oberon, NSW 8, 10-11, 13-17, 34, 47-8, 50-51, 85-6, 89-90, 92-3, 95, 99, 100, 102, 104, 105, 115-35, 145, 146, 148, 149, 150, 152, 154, 169, 173, 175, 177
olives 5, 32-3, 48
One Nation Party 2, 18
Open Days 54, 89, 176
Orange, NSW 24, 34, 36, 100, 101, 126

Pahl, Ray 24, 28
Parkes, NSW xi, 46-7, 51, 63, 70, 72, 81, 90-91, 94, 100, 145, 146, 147, 166

Parramatta, Sydney 68, 81, 107, 109, 111
passport 45-6
pastoral 3-5, 9, 175
peacefulness 42, 73, 105, 107, 124, 127,
 140, 147, 151-5, 175
peri-urban 28, 172, 177
pets 122, 144, 152
place 1, 4-5, 14-6, 18-9, 42, 46, 50, 52, 54,
 56-7, 85, 88, 91-2, 95, 118, 123-6,
 128-30, 133, 135, 171-8
 place marketing/branding 2, 19, 39-40,
 42-84, 85, 89-96, 169-70, 175-7
planning 23, 165, 176
police 17, 50, 56, 79, 121-2, 126, 132, 144
politics 2-3, 5, 18, 20, 40-41, 54, 133-4,
 171-2
pollution 27, 42, 111, 128, 144, 152, 155
population change xiii, 1-14, 16-7, 20-22,
 23, 26-7, 30, 39, 41, 55-6, 58, 61,
 92, 117-8, 163, 169, 178
Port Macquarie, NSW 27, 100
Port Stephens, NSW 157
post-productivity 1, 34, 94, 171
poverty 5, 18, 164
power 19-20, 41, 56
Presley, Elvis 11, 46-7, 63, 68, 81, 90
professionals 30, 45, 55-7, 76, 131, 142-3,
 150-51, 156, 159, 162, 169, 171,
 177
privacy 130, 131, 152, 156, 161
Provence, *see* France
public service 52, 54, 58, 84-5, 117, 121-2,
 125-6, 175
public transport 17, 64, 111, 118, 153, 167
pubs 15, 59, 77, 102, 113, 128, 143, 167

Quilpie, Q 9, 12, 59, 79, 103-4
Quirindi, NSW, *see* Liverpool Plains

racism 83, 112-3, 128
railways 3-4, 17, 77, 118
recession 163, 164
recreation 15, 70, 72, 74, 103, 105, 107,
 131, 135, 144, 146, 159, 162, 166
regions 1-10, 12-19, 21-5, 27, 29, 31-4,
 36-7, 39-46, 50-51, 53-6, 59, 61,
 63-6, 69, 71, 73-5, 78-9, 81-3, 85,
 89-90, 95, 99, 105-6, 108, 110,

 114, 117, 120, 122, 128, 132-5,
 140, 142-4, 147-9, 153-4, 156-8,
 162-5, 168, 172-3, 175-8
regional development 1, 2, 4, 13, 18, 21,
 23, 39-41
religion 120, 145
renovation 25, 131, 140
rentafarmhouse 53, 95, 175
restaurants 35, 59, 66, 69, 82, 90, 131-2,
 143-4, 149, 157, 160-61, 167
restructuring xiv, 14, 28
retirement 1, 17, 19, 24, 25, 26, 29, 46,
 52, 55, 58, 91, 93, 99, 105, 119-21,
 123-6, 131, 133-5, 148, 165, 166,
 175, 177
roads 4, 41, 53, 77, 102-3, 124, 160
Rockhampton, Q 6, 9, 12, 137, 147-50,
 152, 165
Roma, Q 9, 65, 103-5
romance 121, 124
roots 125, 133, 148, 158
Rotary Club 131, 146
Royal Easter Show, Sydney 52, 57
rugby 131, 145
rural and regional Australia 1-6, 10, 17-19,
 21, 39-41, 44, 55-6, 103, 153, 164-
 5, 171, 175
rural idyll xiv, 18, 32-5, 36, 128, 167, 169
rurality xv, 18, 44, 127-8, 130, 167, 173
rural-urban dichotomy 32, 116

safety, *see* security
St George, Q 9, 69
salinity 5, 164
Salt, Bernard 28-9, 34-6, 157, 172, 174
Scandinavia 13, 28, *see also* Norway,
 Sweden
Scone, NSW 4
schools 14, 16-17, 44, 53-4, 56, 65, 68,
 77, 79, 83, 89, 92, 102, 114-15,
 118, 123-5, 132, 140, 142-4, 146-8,
 155, 158-61, 172-3, 176, *see also*
 education, teachers
Scotland 23, 174
sea change xiv, 12, 22, 29, 30-36, 45, 53,
 59, 102-3, 113, 115, 159, 161-2,
 166, 170, 172, 175
Sea Change 30-31, 159

seasons 16, 24, 66, 90, 94, 107, 117, 152, 167
second homes 1, 24, 25, 26, 29, 175
security xiv, 36, 42, 77, 78, 111, 128, 143, 146
self-reliance 3, 27, 116, 123, 175
services 6, 105, 110, 115-6, 117-18, 127, 131-3, 135, 163-4, 165, 171
settlement 3-4, 6, 10, 12, 20, 22, 23, 171-2
sewerage 127
Shoalhaven, NSW 100
Shooters Party 18
shops and shopping 6, 17, 71, 82, 87, 102, 105, 110, 112-14, 118, 126, 127, 129-30, 132, 140, 144-5, 149, 150, 153-4, 159, 161, 163, 166, 172-3
single people 96, 106, 107, 109, 119, 147, 158-9
Singleton, NSW 165
slow food 66, 176
social capital 17, 130
social class 18-19, 27, 32, 39, 43-4, 53, 97, 125, 133-4, 162, 169-71
social deprivation 163
social life 12, 15, 52, 105, 126-135, 145-6
social mobility 135
social networks 106, 129, 134, 158, 160-61, 176
social stigma 83, 163
solar power 123
soldier settlement 4, 23
South Australia 1, 4, 24, 26, 32-3, 40
South Burnett, Q 144, 156, 166
Southern Highlands, NSW 61, 101-2, 122
Spain 133, 134
'spellcheck towns' 172
sponge city 6, 8, 61, 117, 120, 173, 178
sport 14-15, 77, 80, 118, 127, 131-2, 144-6, 151, 161, *see also* individual sports
Stanthorpe, Q 5, 153
suburbs 18, 24, 34-5, 66-7, 72, 82, 97, 107, 111-12, 116, 124, 127, 129, 135, 139-40, 143, 147, 149, 155, 173
success (at Country Week) 21, 39, 43, 48, 58, 61, 63-5, 68, 76, 79, 84-5, 92-5, 97, 137-9, 166, 177-8

suicide 164
Sunshine Coast, Q 12, 26-7
Sweden xiv, 40, 57, 83, 114, 169, 174, 176
Sydney 1-2, 4, 6, 15-6, 19-20, 23, 24, 26, 32, 33, 42, 44, 46, 48, 50, 52, 54-7, 59, 61 89-90, 93, 101, 111, 117-18, 120, 122-4, 126, 128, 132, 139-43, 147, 154-5, 160, 161, 172-4

TAFE (Technical and Further Education) 15, 51
Tamworth, NSW 6, 8-9, 34, 40, 45-6, 51-2, 55, 61, 69, 71, 73, 81, 90, 99-101, 103-4, 111, 141-3, 151, 153-4, 173
Tara Shire, Q 54
Taree, NSW 51, 100, 104
Tasmania 12, 26, 32, 40, 156
teachers 16-7, 52, 56-7, 77, 94, 98, 109-10, 122, 126, 141, 147, 158-9, 162, 164, 172
television 14, 30-31, 33-4, 159, 164, 169-70
Temora, NSW 46, 51, 81, 104
tennis 131
Tenterfield, NSW 40, 61, 76, 81, 103
Thargomindah, Q 9, 46, 50, 65, 72, 173
timber industry 15, 57, 94, 117, 119-21, 126
time 21, 25, 33, 42-3, 48-9, 65-9, 72-4, 78-9, 81, 85-91, 95, 102-3, 107, 111-12, 116, 122, 124-9, 131-2, 134-5, 140, 142-7, 150-55, 158-60, 162, 165-7, 174, 176
Toowoomba, Q 6, 9, 45, 50, 65, 79, 81, 102, 104
Torres Strait 48, 83, 111
tourism 1, 14, 26, 31, 40-41, 43, 50, 61, 85, 89-90, 92, 94, 118, 128, 149, 169, 176
Townsville, Q 154
tradespeople xiv, 15, 98, 108, 150
traffic 124, 128, 143-4, 151, 152, 155
transport 5, 17, 24, 27, 54, 118, 132, 139, 153, *see also* airports, cars, railways, traffic
tree change xiii, xiv, 22, 34-7, 40, 53, 102, 134-5, 155, 170, 172-3, 175, 177-8

trees 35, 48, 127
tropics 26, 73, 152
Tumut, NSW 8, 11, 46, 48, 51, 65, 70-71, 76, 83, 99-100
Tuscany, *see* Italy
Tweed Heads, NSW 27, 100
tyre-kickers 57, 85, 97

umbilical cord 172, 174
unemployment 35, 75, 83, 108, 150, 163
university 15, 40, 51, 66, 81, 105, 147-8, 150
Upper Lachlan Shire 16, 51, 100
Uralla, NSW 40, 54, 61
urban 1-2, 4-5, 8-9, 11-3, 16-22, 39, 41-4, 50, 53, 55, 61, 92, 117, 121, 124, 127-8, 130-31, 133-5, 154-6, 169-78
urban country 24, 27-8, 34-6, 69-71, 72, 73, 82, 133, 135, 153-4, 170, 172, 173-4, *see also* hybrid country, micropolitan
USA 1-2, 12-13, 19, 28, 130

Victoria 4, 15, 26, 30-31, 39-41, 130, 133-4, 158, 177
village 3, 10, 25, 59, 123, 133-4

Wagga Wagga, NSW 6, 14, 46, 48, 59, 61, 77-8, 100, 104
Walcha, NSW 40, 54, 148, 173
Wales 23
Warialda, NSW 45, 50, 61, 81, 100-101, 172-3, *see also* Gwydir
Warrumbungles, *see* Coonabarabran
Warwick, Q 9, 54, 63, 70-71, 93, 104, 149, 152-3, 169
water 2-3, 48, 59, 89, 123, 127, 171, 177
Western Australia (WA) 1, 4, 9, 12, 25, 26, 34
white flight 26, 112-13, 163, 167, 176
Whitlam, Gough 23
wind power 123
wine 5, 25-6, 32-3, 35-6, 48, 59, 61, 69, 71, 129, 157, 171
Wollongong, NSW 1, 16, 20, 61, 161
women, *see also* girls 13-14, 31, 52, 106, 114, 126, 161, 167, 174
work 128, 129, 159

Yass, NSW 46, 51, 69-70, 72-4, 100, 104, 145-6, 152-6, 167, 169
Yeppooon, Q 27
Young, NSW 46, 78
youth 13-16, 125, 178